RECLAIMING THE LAND
Rethinking Superfund Institutions, Methods and Practices

RECLAIMING THE LAND
Rethinking Superfund Institutions, Methods and Practices

Edited by

Gregg P. Macey
University of Virginia Department of Urban and Environmental Planning

and

Jonathan Z. Cannon
University of Virginia School of Law

 Springer

Cover:Photo of Central Chemical Plant, Hagerstown, MD, (fertilizer manufacturing, 1930-1984, EPA Superfund National Priorities List, 1997), 2003.

Brian Gerich, design for 'Watershed Park' (section through remediation biopile and infiltration basins), Community History Workshop and BOOM Studio, Professors Bluestone and Bargmann, 2003. Courtesy of School of Architecture, University of Virginia.

Library of Congress Control Number: 2006935378

ISBN-13: 978-0-387-48856-1   e-ISBN-13: 979-0-387-48857-8
ISBN-10: 0-387-48856-1   e-ISBN-10: 0-387-48857-X

Printed on acid-free paper.

9 8 7 6 5 4 3 2 1

springer.com

TABLE OF CONTENTS

LIST OF FIGURES

LIST OF TABLES

CONTRIBUTORS

Peter Beling, University of Virginia Department of Systems Engineering

Daniel Bluestone, University of Virginia School of Architecture

Todd Bridges, Office of Solid Waste and Emergency Response, U.S. Environmental Protection Agency

Jonathan Cannon, University of Virginia School of Law

E. Franklin Dukes, Institute for Environmental Negotiation, University of Virginia

Alexander Farrell, University of California at Berkeley Energy and Resources Group

Marc Greenberg, Office of Solid Waste and Emergency Response, U.S. Environmental Protection Agency

Jennifer Hernandez, Holland & Knight

Melissa Kenney, Duke University Nicholas School of the Environment

Gregory Kiker, U.S. Army Engineer Research and Development Center

Niall Kirkwood, Harvard University Graduate School of Design

James Lambert, University of Virginia Department of Systems Engineering

Peter Landreth, Holland & Knight

Igor Linkov, Cambridge Environmental, Inc.

Gregg Macey, University of Virginia Department of Urban and Environmental Planning

George Overstreet, University of Virginia McIntire School of Commerce

Faheem A. Rahman, University of Virginia Department of Systems Engineering

F. Kyle Satterstrom, Cambridge Environmental, Inc.

David Slutzky, University of Virginia Department of Urban and Environmental Planning

Mark White, University of Virginia McIntire School of Commerce

ACKNOWLEDGEMENTS

We are grateful to the U.S. Environmental Protection Agency for an assistance agreement under which much of the research reflected in these chapters was carried out. The EPA's generous assistance supported the interdisciplinary Center of Expertise in Superfund Site Recycling at the University of Virginia. Our thanks also to EPA program managers and experts, including John Harris and Melissa Friedland, who offered encouragement and support for the effort, and to the University of Virginia's Jeffrey Plank, who helped bring the Center into being. Our special thanks to the Deputy Directors of the Center, Monique Van Landingham and Caroline Wilkinson, who provided vital coordination and management skills, and to Don Elliott and Karl Gustavson for their helpful suggestions. The book also benefited from the comments and reflections of several anonymous reviewers. A version of Jonathan Cannon's chapter previously appeared in the *New York University Environmental Law Journal* (volume 13). Finally, we would like to thank Bob Irwin, who provided valuable copy editing assistance throughout the manuscript.

FOREWORD

Marianne Horinko
Executive Vice President, Global Environment and Technology Foundation

On August 7, 1978, President Carter declared a state of emergency in the community of Love Canal, New York. The President urged residents of Love Canal to evacuate, not because of a recent catastrophic event, but because of something that occurred in the 1940's and 1950's. This Niagara Falls community had been developed on land that was formerly used as a landfill. Although the landfill was closed in 1953, it had been a dumping ground for tons of chemical wastes, and that waste would eventually create an environment extremely dangerous to human health. The image of chemicals seeping into the basements of American homes would produce widespread panic, but would also raise the environmental consciousness of a nation, and produce a legislative response that was equal to the task.

Americans celebrated the first Earth Day in April, 1970. Throughout the rest of the decade, we passed legislation intended to fulfill the promise of that day: to create a clean and safe environment. However, there were still holes in our environmental protection in 1978, evidenced by the problems at Love Canal. In response, Congress passed the Comprehensive Environmental Response, Compensation, and Liability Act (CERCLA) or, as many people call it, Superfund. Passed in 1980, this law was intended to address problems like the ones faced at Love Canal. It has now been 25 years since its passage and Superfund has proved to be a highly successful program, cleaning up over 900 sites and reducing health risks for tens of thousands of people.

One such site at which Superfund has completed activities is Love Canal itself. In 2004, the site that provided the impetus for passing the Superfund legislation was officially removed from the National Priorities List, signifying that cleanup was complete. As we celebrate the progress over the past quarter century it is also important that we look to the next 25 years with an eye toward not only continuing this progress, but also evolving the program to deal with the challenges we face today.

The removal of Love Canal from the National Priorities List is a useful metaphor for Superfund policy generally. In its early years, Superfund focused exclusively on site assessments and liability apportionment. This response was natural given the circumstances surrounding Superfund's passage. People living near Superfund sites were petrified by the possibility of health problems that these sites could cause and angry at the sites' very existence. This climate was not forward-looking. People were not concerned about site reuse because of the cloud of pessimism that loomed over these sites. However, today is a new day. That era is over. We need to transform Superfund into a more nimble tool, focused on revitalization and capable of adapting as circumstances warrant.

The challenges facing Superfund today are far different than in 1980. The sites like Love Canal – a few acres of buried drums – are largely cleaned up. The sites that remain – the Hudson River, Coeur D'Alene in Idaho, the Fox River in Wisconsin – are much larger and more complex watersheds, defying easy cleanups. Further, Superfund is no longer the only tool in the shed. A wide array of other federal, state, and private cleanup programs have sprung to life. Given limited resources and the need for new models of governance that can adapt to changing intergenerational needs and scientific information, a strategic rethinking of Superfund is in order.

Over the past decade, Superfund has begun the transformation to be more forward-looking. Communities have witnessed these sites be rehabilitated and the pessimism has turned into optimism. People have begun to focus on how to use the sites once cleanup is complete. Even at Love Canal, people are beginning to move back to the homes that were abandoned in the late 1970's. Although we are moving forward, this ethic of optimism must be institutionalized in order to address the challenges of the next 25 years. Everyone has realized that once these sites are cleaned, they represent significant opportunity for communities and developers. If Superfund and the people who administer it acknowledge this reality and operate within a framework where site reuse is an articulated goal, the law will be more effective and the American public better served. This conclusion is not intended to marginalize or obscure the fundamental goal of the program: protecting human health and the environment. It is simply a recognition that the program will achieve these goals more efficiently if plans are seen through the lens of reuse.

At this moment, hundreds of former Superfund sites are being used in every way imaginable. They have become productive once again and add value to their communities in a variety of ways. Site reuse has created tangible, economic benefits in many areas because of increased employment opportunities, property values, and tax revenues, and the potential for additional economic development on surrounding properties. Reuse of these sites provides benefits for their surrounding communities that go beyond monetary values, although the numbers alone are significant. There are many examples of sites being used for recreational activities and as ecological safe havens. Because land earmarked for these purposes is often limited, the reuse of Superfund sites also presents a much-needed opportunity to improve the aesthetic appeal of a community. Although focusing on future use seems obvious, it is an important shift in perspective. We must be mindful throughout the entire process that the site cleanup is directed at rehabilitating the land and returning it to a productive use.

Recognizing that reuse is an important objective from the beginning of the process serves to inform decision-making throughout and improves the overall effectiveness of the program in many ways. First, when site reuse is an explicit component of the cleanup process, we see more constructive community involvement. When people are looking forward to new parks, housing, and shopping centers, they have more reason to find common areas of agreement. Second, we see stronger partnerships between government, private developers, and community organizations, because everyone wins when a neighborhood springs back to life. Third, we see more sensible cleanup plans, because they can be tailored to future uses while ensuring their long-term protectiveness. Fourth, we see easier access to private funding, because cleanup money is seen as an investment with a stream of future returns. Because the partnerships, planning, and funding are targeted at future potential, not past failures, the contamination is often cleaned up more quickly. And fifth, by encouraging sustainable reuses such as green spaces, energy efficient buildings, smart growth community developments, and wetlands, we also are able to prevent the re-contamination of former hazardous waste sites and other indirect environmental problems.

Superfund faces new challenges as it turns 25 years old and we must plan strategically in order to confront them head-on and improve this influential program. We must never compromise the lofty objective of the Superfund, protecting human health and the environment, but we must overhaul our thinking and begin to make this objective compatible with the goal of using the land productively. These goals do not have to be mutually exclusive and, in fact, can work symbiotically in order to maximize resources. However, accomplishing this task will require a fresh outlook on the entire program. Flexibility is necessary. Our approach must evolve with respect to the selection, implementation, and maintenance of remedial measures. As each project is chosen and the process begins, we need to be nimble in how we navigate through the project, while keeping the ultimate goals squarely in focus.

An explicit recognition of the value of reuse is indispensable in creating a more effective Superfund. Integrating planning for future use into the decision making process creates more workable cleanup plans that will not only save money, but will better protect the environment for future generations. Economic realities inform us that redevelopment of cleanup sites present tremendous opportunities for investors. The EPA needs to leverage this knowledge in order to generate private funding and planning that will ultimately expedite the cleanup process while saving taxpayers' dollars. Incorporating future use into the evaluation process will allow the EPA to achieve these benefits while building in safeguards to ensure that its role as protector of human health and the environment are never compromised.

In addition to integrating future use into its decision making process, the EPA must also take steps to remove unnecessary and excessive impediments to site reuse. A critical and transparent analysis of what worked and what did not will allow the EPA to move forward in a manner that more effectively achieves all community objectives. Being hamstrung by technical minutia is not a problem for Superfund administration alone. It is a prevailing problem in implementing many statutory mandates, especially in the environmental context. We must focus on the problems we face in a more holistic manner. Whether it be narrow statutory interpretations or a lack of innovative thinking, the EPA must break down barriers and begin to see the forest for the trees when it concerns site cleanup and reuse.

I want to underscore the fact that the EPA's primary responsibility is to protect human health and the environment and the future of Superfund must advance this objective as it always has. There must be monitoring mechanisms in place that allow for real oversight so that site use remains protective and land use controls are adhered to. However, there must also be flexibility inherent in the process so that the EPA can make informed decisions to modify directives so that the process respects changing land use patterns and community needs. Undertaking cleanups at many of these sites is incredibly complex. Issuing an inflexible set of mandates to govern the entire cleanup process is at best inefficient and at worst unworkable. Where there are private parties willing to contribute resources to the cleanup effort, the EPA needs the agility to rethink its cleanups or components of those cleanups. This agility will lead directly to quicker community revitalization while placing less strain on public funds.

Superfund is entering a new era that is both exciting and daunting. The last decade has produced much progress towards a holistic approach to site cleanup and reuse. However, the job is unfinished. In order to fulfill Superfund's promise for the next 25 years, we must look critically at our past and plan strategically for our future. The authors of this book understand this reality. The book provides the reader with a set of innovative management approaches for the new era of Superfund in which site reuse is embraced. The book seeks to tackle the greatest challenge facing Superfund in general, and reuse in particular, which is the intrinsic uncertainty involved in complex site cleanup and redevelopment. Building on the concepts of adaptive management, the book offers a vision of Superfund that is both flexible and dynamic. It envisions a program ready to respond to changes in site cleanup, while always keeping future use options in perspective. The real value of the adaptive management approach is that it provides the flexibility to deal with any uncertainty that arises. For example, the authors demonstrate how stakeholder interests can change throughout the cleanup process and how adaptive management provides the proper approach to handle that situation and keep the project moving forward while adapting to the new circumstances.

Rethinking the broader Superfund strategy is a critical step in transforming this program into an effective tool for the future. The ultimate goal of site reuse must inform and influence the results of this transformation. The lens of site reuse allows us to focus on broader community concerns, while maximizing the effectiveness of public and private resources. The reuse of sites represents a move beyond the singular, essential goal of protecting human health and the environment, and embraces the increasing importance of land as a source and a resource for community revitalization. It is a process of theoretical and tangible integration. We will conceptualize the site as ultimately contributing to the community in order to inform the process. Additionally, we will work efficiently to physically reintegrate the site into the community. The era of leaving sites fenced off from public use for decades at a time is over. We must plan for the future, stabilize and secure the safety of the site, and utilize the resultant resource to improve the community. This project is inherently optimistic but fundamentally realistic, and it can be achieved through a simple strategic change.

This book fills an important void in the literature surrounding Superfund and will facilitate a necessary public discourse on how to transform that program so it is capable of effectively addressing the problems of the next 25 years. Both timely and insightful, this book outlines a sensible approach for the future of the program that promises to streamline implementation and transform Superfund into an optimistic tool capable of drastically improving multiple aspects of our communities and our lives.

INTRODUCTION:

The Promises and Pitfalls of Adaptive Site Stewardship

Gregg P. Macey

University of Virginia Department of Urban and Environmental Planning

In the late 1970's, residents of Niagara Falls, New York began to experience the effects of decisions made by one company three decades earlier. The Hooker Chemical Company dumped over 21,000 tons of chemicals, including dioxin, into a nearby canal. No one had reason to suspect that the dumping occurred until one day, the substances began to seep into the basements of homes and schools. The public outcry that followed the Love Canal incident led to passage of the most advanced hazardous waste cleanup program in the world, which to date has generated more than twelve billion dollars in commitments to remedy affected sites. Known as the Comprehensive Environmental Response, Compensation, and Liability Act (CERCLA, or "Superfund," after its funding mechanisms), the legislation gave the Environmental Protection Agency (EPA) the authority to clean up contamination from past disposal practices that are found to pose risks to human health or the environment.

Such a vast federal program, already many times larger than originally intended, still pales in comparison to the scope of the problem posed by contaminated properties. Roughly one in four Americans, including ten million children, lives within four miles of a toxic waste dump. And while estimates vary, at least 200,000, and probably more than 500,000 sites (sometimes referred to as "brownfields") in the United States contain either soil or groundwater that may require remediation to overcome the negative effects of past industrial operations. Many of these are ideally located in industrial or commercial zones close to urban centers and essential infrastructure, yet they continue to sit idle, contributing to urban blight and limiting job creation as well as tax revenues for local governments. These properties do not even include the large expanses of land operated by the Department of Defense and Department of Energy. For example, the DOE spends between $5.6 and 7.2 billion per year on the environmental management of its sites.

COEUR D'ALENE: A THIRD GENERATION CERCLA CHALLENGE

Estimates of the number of contaminated properties in the United States also leave out vast geographies that call for complex cleanup efforts, including the Hudson River, New Bedford Harbor, and the Palo Verde Shelf in the Pacific Ocean offshore from Los Angeles. These megasites exacerbate the challenges posed by contemporary contaminated lands, including the uncertainties surrounding risks posed, effectiveness of cleanup options, and complex social, economic, and political settings in which they exist. One such area that gives us a sense of the magnitude of these challenges was formerly one of the leading silver, lead, and zinc-producing regions in the world. The Coeur d'Alene (CDA) Superfund site, spanning the states of Washington, Idaho, and Montana, was added to the National Priorities List in 1983. It is an immense landscape, incorporating "mining-contaminated areas...adjacent floodplains, downstream waterbodies, tributaries, and fill areas" near the Coeur d'Alene River corridor, in addition to a 21-square mile region of historic smelting operations known as the "Bunker Hill Box" (EPA 2002: 1-1). As is standard practice, the site was divided into different geographic units by the EPA: an Upper Basin (former and current mining, milling, and processing), Lower Basin (the river itself, adjacent lateral lakes, the floodplain, and associated wetlands), Coeur d'Alene Lake, and depositional areas of the Spokane River. Three operable units (OU's) were identified in order to facilitate management of the range of remediation challenges facing the region: (a) populated areas of the Bunker Hill Box, (b) non-populated areas of the Box, and (c) mining-related contamination in the river basin. Contaminants that threatened human health (metals such as lead and arsenic) and ecological receptors (lead, cadmium, and zinc) were identified, remnants of the nearly 900 mining and milling-related features in the region that over the years yielded 1.2 billion ounces of silver, 8 million tons of lead, and 3.2 million tons of zinc. An operation of such scale naturally resulted in a broad range of waste products, including tailings (parts of the ore from which metals could not be recovered economically), waste rock (non-ore rock that had been taken from a mine), and smelter emissions. The EPA estimated that the total mass of impacted materials was more than "100 million tons dispersed over thousands of acres" (EPA 2002: 2-5).

EPA documents demonstrate the enormity of the remediation challenge presented by such a large-scale operation: more than ten years after regulatory actions began on the site, the EPA reported that "given the extensive contamination present, the bulk of the mining-related wastes that are deposited throughout the river and floodplain still remain" (EPA 2002: 2-6). The speed of remediation efforts masks an even more fundamental challenge posed by such a site: the pervasive uncertainty under which agency officials must operate and make decisions for the region. For example, the first study of household effects of mining-related contaminants (particularly lead absorption by children) to take place outside the immediate area of the Box began in 1996. The Idaho Department of Health and Welfare and the Agency for Toxic Substances Disease Registry (ATSDR) collected a variety of samples in 1997: soil, sediment, groundwater, surface water, indoor dust, lead-based paint, and garden produce throughout the CDA Basin. An RI/FS (remedial investigation/feasibility study) process commenced the following year, including field sampling and a quality assurance plan. The sampling occurred in waves, including sampling plans developed as "field sampling plan addenda to the base plan" to address data gaps found after "reviewing available historical data and results of previous field sampling" (EPA 2002: 2-6). Over ten thousand samples were collected. Yet in 2002, the EPA concluded that

> [T]he large geographic area of the Basin made it impractical to collect all the data needed to fully characterize each source area or watershed. Further data collection will be necessary to support remedial design for areas identified as requiring cleanup. This may include areas where previous cleanup actions have taken place, such as floodplain areas of the Union Pacific Railroad right-of-way or other areas where previous removal actions have addressed some, but not all, contamination present (EPA 2002: 2-7).

The extent of the public health threat remained partially obscured even after such interventions as a Lead Health Intervention Program by the Centers for Disease Control and ATSDR, a time-critical removal action of sixteen public properties (such as city parks and school playgrounds), and a yard soil removal program. Agencies working on such efforts learned that they would have to remediate "at least 200 residential yards each year" in order to avoid recontamination from parcels that had yet to be cleaned up (EPA 2002: 4-1). The parties involved in cleanup efforts benefited from five-year reviews of the individual operable units, and adjusted their activities accordingly. For instance, the EPA determined that house dust would have to undergo extensive sampling "should house dust lead levels remain elevated following completion of yard soil remediation" (EPA 2002: 4-2).

Across the OU's, federal and state agencies applied numerous cleanup tools at their disposal: source removals, surface capping, surface water creek reconstruction, milling and processing facility demolition, closures for waste consolidated on site, revegetation efforts, and surface and groundwater controls and wetlands treatment. But the limits to comprehensive site characterization and cleanup emerged in OU's focused on human health and ecological receptors alike. OU 2 consisted of non-populated areas including the former industrial complex, mine operations areas, and the Bunker Hill Mine. Smelter stabilization work began in 1989 and ended in 1993. After this, potentially responsible parties signed a consent decree to conduct cleanup activities in the area and the EPA and the State of Idaho entered a State Superfund Contract to carry out remaining site remedial actions in 1995. A five year review of OU 2 followed in 2000. While efforts to consolidate contamination from various areas of the site were completed by 2001, the ROD for OU 2 failed to indicate a response action for mine water:

> The ROD, therefore, did not address control of acid mine drainage from the Bunker Hill Mine or operation of the Central Treatment Plant (CTP) in any significant way. The ROD briefly addressed the mine water by requiring that it continue to be treated in the CTP prior to discharge to a wetlands treatment system for removal of residual metals. During studies conducted between 1994 and 1996 by the US Bureau of Mines, the wetlands treatment system was found to be incapable of meeting the treatment levels established in the ROD. The 1992 ROD did not contain or otherwise identify any plans for the control or long-term management of the mine water flows. The ROD also did not address the long-term management of treatment residuals (sludge) from the CTP, which are currently pumped into an unlined pond on the CIA. At the current disposal rates it is estimated that the pond will be filled in 3 to 5 years (EPA 2002: 4-3).

Such challenges as dealing with treatment residuals were addressed in amendments to the OU 2 ROD, illustrating the kind of learning and adjustment that the five year review process encourages. The true limits to the EPA's ability to address the scope of remediation challenges posed by the CDA site, however, emerged during implementation of the ROD for OU 3, which focused on the CDA Basin itself. The EPA joined the State of Idaho, Coeur d'Alene Tribe, and other federal, state, and local agencies to form the CDA Basin Restoration Project, an initiative that the EPA concluded "had limited success as a systematic approach to addressing contamination in the Basin" (EPA 2002: 4-4). The scope of the challenge for OU 3 was immense: develop a water quality improvement program for the Basin by coordinating regulatory authorities under the Clean Water Act, CERCLA, RCRA, and other programs, an effort that would have to address non-discrete sources as "the primary sources of metals in surface water in the Basin" (EPA 2002: 4-4). Even the 2002 ROD for OU 3 recognized exposure pathways that remained unaddressed, including recreational use areas in the upper and lower basin, subsistence fishing by the CDA and Spokane Tribes, and potential future use of groundwater that remained contaminated with metals. The Selected Remedy for OU 3 identified "thirty years of prioritized actions in areas of the Basin upstream of the CDA lake" (EPA 2002: 4-5). The EPA was keenly aware of its limited ability to forecast how its efforts would impact the region:

> EPA expressly recognizes that after the selected remedial actions are implemented, conditions in the Upper and Lower Basin may differ substantially from EPA's current forecast of those future conditions, which is solely based on present knowledge. The tremendous amount of additional knowledge that will be gained by the end of this period through long-term monitoring and five-year review processes may provide bases for future Applicable or Relevant and Appropriate Requirement (ARAR) waivers. In addition, this new information and advances in science and technology may allow for additional actions to achieve ARARs and protect human health and the environment in a more cost-effective manner (EPA 2002: 4-5-4-6).

The EPA reiterated the incompleteness of its actions under any given ROD in its discussion of OU 3:

> The remedial actions selected in this ROD are not intended to fully address contamination within the Basin. Thus, achieving certain water quality standards developed under the Clean Water Act and the Safe Drinking Water Act, such as water quality standards and Maximum Contaminant Levels, are outside of the scope of the remedial action selected in this ROD and are not applicable or relevant and appropriate at this time. Similarly, special status species protection requirements under the Migratory Bird Treaty Act and Endangered Species Act are only applicable or relevant and appropriate as they apply to the remedial actions included within the scope of the Selected Remedy...Because this remedy will result in hazardous substances, pollutants, or contaminants remaining on-site above levels that allow for unlimited use and unrestricted exposure, statutory reviews will be conducted at least every five years after initiation of remedial action to ensure that the Selected Remedy is, or will be, protective of human health and the environment (EPA 2002: 13-2, 13-20).

Possessed with insufficient information "to characterize all the specific sources of metals contamination impacting the streams and floodplains, as well as the anticipated effectiveness of certain remedial actions," the EPA designed its initial set of actions to take place in "defined locations" and to achieve "specific benchmarks." But the range of possible actions that might be called for as the Selected Remedy for OU 3 progressed, and the scope of uncertainty facing parties to the cleanup process remained. The estimated present value cost of the Selected Remedy for OU 3 was $359 million in 2002, with cost accuracy listed as between -30 and +50% of that value. The size and complexity of the site called for a new approach to remediation, which the EPA referred to as "an adaptive management strategy to implement cleanup."

Whether the EPA's use of an "adaptive management strategy" for the CDA site offered a unique approach to dealing with environmental contamination on a large scale, which posed an array of human and ecological health risks and demanded the attention of numerous agencies and a timeline that would span decades, is unclear from EPA documents. Immediately following the EPA's use of the term "adaptive management," the ROD for OU 3 lapses into a standard depiction of how the agency will deal with uncertainty: a selected remedy included prioritized actions for certain areas, a remedial investigation will collect data needed to characterize the site, remedial alternatives will be developed and evaluated, a remedy will be selected, and additional data will be collected to support the design of the remedy. Indeed, the EPA characterizes this process of anticipating the need for additional design data as "not unique to the CDA Basin." The centrality of long-term effectiveness and monitoring within the choice and evaluation of a Selected Remedy, while encouraging, relied less on a complete reordering of Superfund tools to meet a unique challenge posed by large-scale sites and more on the selective use of some of the more immediately accessible elements of adaptive management.

Such was the conclusion of the National Research Council in an exhaustive study of the CDA Superfund Site's remediation objectives and approaches to date, a document that ends with the recommendation that adaptive management be "unequivocally incorporated into every step of the Superfund process, beginning with the Remedial Investigation (NRC 2005: 273). The report focused on the difficulties facing "megasites" such as the CDA River Basin. Their geographic scope, volume and complexity of contaminated material, lack of obvious engineering solutions, and unthinkably long time horizons (which for natural recovery and achievement of ARAR's were projected "up to 1,000 years" for the CDA site) figured prominently in the report (NRC 2005: 264). The need to manage megasites under conditions of uncertainty was a common motif repeated throughout the NRC's findings. Examples included the lack of a "definitively identified" causal link between remediated yards and decreased blood lead levels, the presence of "floods and other actions" that "eroded the installed remedies or caused recontamination" in the Box, the need for "institutional mechanisms to monitor…effectiveness, repair any failures, and remain in place and effective for an extremely long time (at least hundreds of years)," and the lack of remedies "to address risks from possible future uses of contaminated groundwater and risks to residents who engage in subsistence lifestyles" (NRC 2005: 262-263). In the face of such challenges, the present architecture of the Superfund program as summarized by the NRC appears strikingly outmoded and overmatched:

> The Superfund process calls for EPA first to gather all the necessary information (the remedial investigation phase), then evaluate alternatives for addressing all the human health and environmental risks identified in the information-gathering stage (the feasibility study stage), and then decide on the best remedies for reducing these risks to acceptable levels (the Record of Decision)...At most sites, the OU being assessed addresses only one or two closely related problems, and this process works reasonably well. In the CDA OU-3, however, there are a large number of different problems. Some, like the contamination of yards, are fairly easy to assess. Others, like the reduction of dissolved metals in the main stem of the river are much more difficult. By combining these different problems into one OU and subjecting them to the process established in the NCP, EPA must attempt to answer all the questions for all the problems before it can attempt to remedy any of them (NRC 2005: 317).

The NRC ended its discussion of the Superfund program by noting that whatever flexibility could be found within CERCLA and the NCP "does not appear sufficient to address all the issues identified by the committee" (NRC 2005: 320).

MEETING THE CHALLENGE

The realities unearthed by the NRC report call upon the Superfund program to evolve in a way that accommodates new scales of operation, dealing with the intricacies of large-scale projects while not posing obstacles to redevelopment of smaller parcels of land such as brownfields. There has yet to emerge a cohesive strategy for doing this. While the Superfund program (in conjunction with other legislation designed to track hazardous chemicals from production to disposal) has generated some clear success stories, its focus on the liability of those who benefited from improper disposal of hazardous waste is considered the major obstacle to redeveloping brownfields. At the same time, the program as it was created emphasized rapid and complete cleanups, operating under the assumption that contaminated sites can be dealt with effectively over a relatively short period of time. Such an approach leaves the EPA and other parties ill-prepared for the challenges posed by sites that, like Love Canal, are only brought to federal attention after decades of gradual or intermittent contamination, or by potential Superfund sites that, either because of their size or characteristics, are themselves a part of larger ecosystems that require more than short-term intervention.

Reauthorization legislation, three rounds of reform in the mid-1990's, and passage of the Brownfields Revitalization Act of 2002 have left the essential architecture of the program intact and inappropriately designed for the next generation of environmental challenges and redevelopment needs posed by contaminated properties. To date, there has been a tendency of state and federal programs to try to remove barriers to the redevelopment of contaminated sites one at a time: incentives for voluntary cleanup are enacted, grants are offered for site assessment, and technical assistance is given to developers, among other efforts. When faced with large-scale contamination, Superfund allows for some flexibility in addressing the uncertainties posed by interventions within a watershed, mountain ridge, or densely populated urban center: sites can be broken into operable units and addressed separately, and interim remedies are sometimes authorized. These piecemeal tactics do not offer agency officials, responsible parties, or affected communities a lens through which to view and evaluate the program's ability to cope with the uncertainties posed by large contaminated sites. They attempt, through such innovations as prospective purchaser agreements and pollution liability insurance, to bring some level of certainty to the process so that potential landowners are able to pursue their redevelopment goals. What they neglect is the inherent *uncertainty* involved in complex site cleanup and redevelopment, a reality that calls for a fundamentally different framework and approach.

This book offers such a lens. Its authors, many of whom participated in the Center for Expertise for Superfund Site Recycling (based at the University of Virginia), have for the last three years considered new approaches to revitalizing contaminated land. Their approach is interdisciplinary, combining advanced quantitative methods for site characterization and optimization with qualitative tools for site visualization and design and community participation. These techniques were developed with an understanding that contaminated sites should be treated not as isolated spaces to be defined by regulatory boundaries and addressed one medium (water, soil) at a time, but as part of broader ecosystems, which call for site interventions in multiple phases, over extended periods of time, and under conditions of uncertainty and change. In this spirit, our book, *Reclaiming the Land*, builds upon thirty years of lessons learned by an approach to natural resource management known as adaptive management.

THE ADAPTIVE MANAGEMENT APPROACH

Adaptive management established itself as an alternative approach to natural resource management in the 1970's (Holling 1978). Lee, who popularized the approach in the literature, defines adaptive management as "treating economic uses of nature as experiments, so that we may learn efficiently from experience" (Lessard 1998).[1] His book, *Compass and Gyroscope*, is the most widely cited treatment of the subject in the literature (Lee 1993). Lee's work is instructive in that it focuses on the organizational dimension of "learning while doing" as a complement to more technical approaches. He declared that "the field is profoundly different from a laboratory" (Lee 1999: 3), a reality that calls upon natural and social scientists to make use of new and different modes of learning if they are to effectively manage at the ecosystem level. Some challenges to learning posed by ecosystem-level management include:

1) Causal understanding emerges at a slower rate than the efforts to understand them;
2) Usually it is not possible for scientists to separate the effects of management decisions from those of concurrent shifts in the natural environment; and
3) Problems are identified at moments of crisis (such as when there is a rapid decline in the vitality of a given species); these extreme moments tend to be followed by less extreme periods, giving the illusion that management decisions were successful (Lee 1999).

Effective experimentation occurs when the limits of learning resulting from human interaction at the ecosystem level are addressed in the design of managing institutions. "A key unanswered question," therefore, "is whether the adaptive capacity of both ecological and social systems can keep pace" with the expanding influence of human constructs on their environment (Prichard et al. 2000: 38). The central project of adaptive management is to develop flexible institutions that are attuned to ecosystem dynamics (Folke et al. 1998). Where policymakers fall short in meeting this objective, it is often

---

[1] McDaniels and Gregory (2004) added that "Adaptive management proceeds from the premise that policies can be treated as experiments. It involves trying different policy actions in informative contexts, creating experimental designs with controls where possible, avoiding costly failures, monitoring and evaluating outcomes, and selecting a basis for judging what has been learned." The National Research Council (1994) defines "the core adaptive management experiment" as including the involvement of stakeholders in a collaborative process, development of a vision, creating a mission statement, and setting measurable management objectives and informational needs.

due to flaws in institutional design,[2] problems with governance (Ostrom 1990), or the pathology of resource management. The latter involves a process where policy decisions prove successful, leading to myopic research and management behavior that eventually reduces ecosystem resilience (Holling & Meffe 1996).

True to the common conception of adaptive management, a growing number of management experiments have been developed to try to modify fisheries regulations, balance hydropower uses, restore riparian habitat, balance agricultural and other uses, develop storm water management plans, and improve the implementation of Habitat Conservation Plans, among other applications (Freedman et al. 2004). These efforts have led researchers to recast adaptive management's experimental focus as part of a broader framework for dealing with the uncertainty and longer time horizons that confront environmental managers who function at the level of ecosystems. For example, Torrell (2000: 354) defines adaptive management as designed "to cope with the uncertainty and complexity of ecosystems by creating spaces in which reflection and learning can occur and by allowing management processes to take action in light of new information." His approach treats experimentation as one of three dimensions, the others being project strategy adjustments as new information is obtained and active participation by relevant stakeholders.

The participatory dimension of adaptive management is in keeping with Lee's assertion that political conflict and dispute resolution can provide means of error recognition, which is a necessary complement to the kind of "self-conscious learning" to which adaptive management aspires (Lee 1993: 87). Negotiation and planning are the means through which policy-oriented learning is accomplished.

> Because control over large ecosystems is fragmented, the search for sustainability requires extensive social interaction: sharing analytical information such as simulation models and databases, identifying tradeoffs and coalitions for joint action, and learning from surprising outcomes. These interactions become ways to

---

[2] Several institutional conditions can encourage adaptive management: (a) a mandate to take action in the face of uncertainty, (b) awareness among decisionmakers that they are experimenting, (c) care taken to improve outcomes over biological time scales, (d) understanding that human intervention cannot produce desired outcomes predictably, (e) sufficient resources to measure ecosystem-scale behavior, (f) availability of theory, models, and field methods to estimate ecosystem-scale behavior, (g) the ability to formulate hypotheses, (h) an organizational culture that encourages learning from experience, and (i) sufficient stability and institutional patience to measure long-term outcomes (Lee 1993: 63).

negotiate shared substantive agendas that individual organizations and interests cannot achieve by themselves...Once a framework for conducting the dispute is in place, it is possible for parties to undertake the substantive task of planning – the assembly of information and analytical skills to describe the world shared by the parties and to identify the uncertain consequences of action within it (Lee 1991: 785, 787).

While this aspect of adaptive management has received little formal treatment in the literature (McDaniels & Gregory (2004: 1921) note that "adaptive management has sometimes floundered because of inattention to concepts of good collective decision-making with stakeholders, while stakeholder processes have often neglected the importance of learning and adaptation"), the challenge is clear: take public involvement, which is often relegated to a discrete event in environmental decision-making ("a snapshot of pre-project conditions"), and reconstruct it as a dynamic process that is cognizant of changes over time (Shepherd & Bowler 1997). The move toward viewing Superfund sites as ecosystems, the shift in objectives from remediation to reuse to long-term stewardship, and the need to involve residents who are seeking to reclaim their communities, build narratives around abandoned sites, and question assumptions underlying the models and mindsets of government agencies, speak to the relevance of the adaptive management framework, particularly its neglected emphasis on participatory process.

A well-developed treatment of adaptive management sets experimentalism next to two additional core principles of the approach: multi-scalar analysis ("adaptive managers model and monitor natural systems on multiple scales of space and time") and place sensitivity ("adaptive managers adopt local places, understood as humanly occupied geographic places, as the perspective from which multi-scalar management orients") (Norton & Steinemann (2001: 477). Scholars view these principles as operating within a nonlinear process including several commonly used steps, such as monitoring, evaluation, and feedback. For instance, Lessard (1998: 81) outlines the following components of the approach:

(1) **Assessment**: understanding the current ecological conditions of places of interest, changes in ecosystem components over time, and likely trends within them; developing scale-relevant assessments that place social, economic, biological, chemical, and physical components of a management area into a larger ecological context;

(2) **Scenario planning**: identifying "critical uncertainties," obtaining the best information on them, and designing a monitoring and evaluation system to track decisions; this component is cognizant of the trade-off between designs aimed at avoiding failure and those that respond and survive in the face of failure;

(3) **Goals and objectives**:   using assessment to assign values to current conditions and describe desired future ecological conditions; efforts to encourage public ownership of this step are critical;

(4) **Hypothesis development**:   creating an experimental design and readying it for implementation; focus is on identifying techniques that reduce uncertainty and benefit from it; and

(5) **Monitoring and evaluation**:   determining what information should lead to changes in policy; new information is gathered from a broad range of sources, including monitoring, regulatory shifts, and organizational assessments.

Perhaps the easiest way to define adaptive management is through an understanding of its alternatives: trial-and-error learning and deferred action. Trial-and-error learning pervades much of natural resource management, emphasizing resource use while ignoring error detection (which requires costly monitoring systems). Trial-and-error has been implicated in the kind of reactive learning that characterizes the pathology of resource management (Wilhere 2002). Walters (1986) first described the pattern of crisis and opportunity that occurs in resource systems, a pattern of ecological surprise and policy response that has been identified at various scales of resource as well as bureaucratic systems (Johnson et al. 1999). Deferred action means that attempts at ecosystem management are postponed until after the system is understood. By contrast, adaptive management acknowledges the need for regular adjustments to policies and designs them so that change is internally driven.

PROBLEMS AND PROMISES FOR ADOPTING THE APPROACH

While few have questioned the need to move from deferred action and trial-and-error learning to experimental interactions with nature (Lee & Lawrence 1986), the limitations to an experimental approach to resource stewardship have been well-documented. A broad-based criticism of adaptive management is that the bulk of the literature "has not moved far beyond individual, place-based experiences – most experiments in adaptive management work in isolation, with little interaction, sharing of lessons learned, or comparative assessments of how various efforts are contributing to the advancement of knowledge" (Light 2002: 43). Indeed, a plethora of $n = 1$ case studies dominates the literature, serving a useful role but including little comparative or statistical analysis.

Perhaps of greater concern are the "perverse incentives" that adaptive management programs often encourage. Ascher (2001) noted that in order to implement adaptive management principles,

> [T]he relevant officials need to be responsible for a broad range
> of functions: foresight, sophisticated doctrine formulation,
> effective implementation, learning, and adaptation. Government
> agencies must be able first to identify policy weaknesses and
> then adjust to overcome these weaknesses. They also must be
> able to recognize which aspects of resource control are best left
> to nongovernmental actors and then exercise self-restraint to
> leave those aspects in the hands of these actors.

Government agencies and officials often do not meet such challenges, because the complexity of ecosystem management either "threatens the institutional interests of these agencies or provides opportunities to enhance their interests" (Ascher 2001). Ascher outlines the characteristics of adaptive management that are most prone to perverse incentives, including oversimplification in the face of complexity, the persistence of established organizations designed for resource management, and time horizons that reflect institutional interests rather than areas of concern. They are reminiscent of the work of organizational theorists such as Meyer and Rowan (1977) and March and Simon (1958), who helped to define the ceremonial qualities of formal organizational structures that are often at odds with the preferences of internal actors or pressures exerted by the environment in which they function.

Another blind spot in the literature is what Moir and Block (2001: 141) term adaptive management's "weakest link": the information feedback system. While they acknowledge that a failed adaptive management approach may be attributed to a variety of factors (such as agency resistance to change and poorly defined monitoring), Moir and Block are more concerned about adaptive management's claim that "activity will be modified or stopped at the earliest sign of adverse impact or when it becomes clear that there is significant divergence from the trajectory towards stated goals" (142). Knowing when there is truly an adverse impact or divergence is an art and a science riddled with challenges: monitoring systems are usually scaled to the near-term, managers do not notice warning signs that emerge as part of slower-moving processes (Holling 1995), and early responses to interventions can be both noisy and fleeting while certain thresholds or latency periods remain hidden. To correct this shortcoming, the authors note that managers must be able to "understand the frequency and amplitude of the processes and functions of the ecosystem," "recognize the slower, longer cycles in ecosystem dynamics and design the monitoring frequency to accommodate those cycles," and

> [C]onsider what happens when extreme events occur; when there
> are high levels of noise; when and if complex, nonlinear, highly
> interactive ecosystem processes converge at some critical point,

and when ecological thresholds might be exceeded in the long-term (146).

Faced with challenges such as avoiding perverse learning and identifying new information or thresholds that would compel adjustment, the design of experimental management programs continues. The collective experience of researchers in the field of adaptive management offers a rich set of ideas for those involved in the next generation of Superfund law and practice, which attempts to view contaminated sites on multiple scales over extended time horizons and with new roles and responsibilities for citizens, developers, agencies, and responsible parties. It also suggests several roadblocks in the path toward focusing on ecological systems that cross administrative boundaries, applying systems dynamics and approaches that are attuned to matters of scale and ensuring ecological resilience (Pritchard et al. 2000), and encouraging broad-based involvement in implementation.

The first lies in statutory and organizational resistance to change within agencies, other stakeholder groups, and environmental management as a field. Walters (1997) noted that because adaptive management policies result primarily from court decisions and legislative acts, and because most funding comes from such acts, "sculpting the legislation in a manner that dictates adaptive management philosophies" is key. One can find numerous statutes where adaptive management is called for to some extent. For example, Freedman et al. (2004) show how the Clean Water Act includes a number of adaptive processes rather than definitive determinations (such as NPDES permit reevaluation over five-year cycles, triennial review of state water quality standards, multiyear cycles for state water quality assessments, and continuing planning processes throughout). They take the conventional framework for Total Maximum Daily Loads and convert it to an adaptive process, where managers begin with preliminary allocations and progressively improve controls as their understanding of a water system improves. This frees stakeholders from time-consuming and often irresolvable disputes over uncertainty of numeric values and whether "final" recommendations are optimal. Similarly, Phillips and Randolph (2000) compare goals described in the National Environmental Policy Act with ecosystem management principles. They suggest that an ecosystem management approach could provide the means for agencies to get beyond the dominance of procedural requirements embodied in Section 102(2)(c) of the Act. Steyer and Llewellyn (2000) provide a thoughtful account of regulations designed to slow coastal land loss through passage of Act 6 by the Louisiana Department of Natural Resources. Adaptive management principles were embedded within the programs that resulted.

The EPA and other agencies will at times implement programs using a logic that is opposed to the principles of adaptive management. Landy,

Roberts, and Thomas (1994) discuss this logic in *The Environmental Protection Agency*: *Asking the Wrong Questions*. They describe how the EPA, since its creation, has been reluctant to deal with matters of uncertainty, pushing them to the side and functioning as though ecosystems contained well-defined distinctions that can be identified and linked to words used in regulations (for example, the suggestion that there is such a thing as a "most sensitive group" in human settlements, while the real world suggests a continuum of sensitivity). The statutory constraints to implementing adaptive management principles for Superfund site stewardship that are imposed by CERCLA are one focus of Chapter 2, "Adaptive Management in Superfund: Thinking Like a Contaminated Site."

Chapter 2 also addresses the institutional barriers to effective use of adaptive management in the Superfund program, an approach that its author argues involves recharacterizing Superfund sites as part of the broader ecosystems in which they exist. The first step toward removing institutional barriers demands a realization that organizational decision-making and learning does not always proceed according to rational choice, a fact that was assumed in the design of many of our environmental policies. Lee (1993: 138-139) highlighted recommendations made by environmental policies that assume a rational learning process. NEPA calls upon planners to protect against foreseeable serious hazards, the Endangered Species Act assumes that managers will be able to rely on past experience to determine which species are endangered, pesticides regulations express confidence that laboratory experiments can reduce uncertainty, and nuclear regulators operate under the assumption that the most "important" hazards can be identified and addressed through the redirection of resources. Rational choice assumes that individual actors in organizations update their understanding, modify choices as new information improves comprehension, view competing goals together, and make tradeoffs, compromises, and creative solutions that integrate conflicting values into a coherent set of goals.

In contrast, organizational theorists would argue, starting with March and Olsen (1989), that decision-making is often contextual, infused with ritual, and concerned as much with interpretation as it is with specific choices. This realization has led theorists, such as Weick (1993), to shift the focus in part from decision-making to meaning, investigating how organizations create order and "retrospective sense" of situations. While decision-making seeks to remove ignorance, sensemaking tries to reduce confusion. The latter strives for resilience in the work groups, agencies, and corporations that face crises or the need for change. Improvisation, understanding that ignorance and knowledge grow together (and that confidence and caution eliminate curiosity, openness, and complex sensing, which are most needed for change), and moving away from attempts to make role systems change fast enough to keep up with changing environments can each lead to increased adaptability of organizations. These approaches are indicative of the institutional arrangements needed to implement and adjust ecosystem-based management programs (Imperial 1999a, 1999b): those that address the challenges posed by "bounded rationality"[3] and "cybernetic"[4] and "single-loop" learning,[5] which lead organizations away from learning and limit relationships between individual-level learning and changes in institutional behavior.

As statutory and institutional resistance to change is reduced, and new approaches to Superfund site stewardship explored, we will face an enormous challenge when we begin to implementing policy changes at specific sites: crafting methods and techniques that allow for management at multiple scales and across longer time horizons without crowding out residents, their experiences and goals, and the comparative advantages that they enjoy over project managers and technocrats. Chapters 3-5 of *Reclaiming the Land* focus on this element of Superfund process reform. These chapters are derived in part from the dominant model of policy analysis, based on systems analysis,

---

[3] Bounded rationality involves decision-making aimed at satisfactory rather than optimal outcomes. It recognizes that usually individuals cannot identify or understand what would be an optimal choice:

> The administrator recognizes that the perceived world is a drastically simplified model of the buzzing, blooming confusion that constitutes the real world. The administrator treats situations as only loosely connected with each other – most of the facts of the real world have no great relevance to any single situation and the most significant chains of causes and consequences are short and simple. One can leave out of account those aspects of reality – and that means *most* aspects – that appear irrelevant at a given time. Administrators…take into account just a few of the factors of the situation regarded as most relevant and crucial. In particular, they deal with one or a few problems at a time, because the limits on attention simply don't permit everything to be attended to at once (Simon 1997: 119).

Lee noted that rapidly changing environments "pose particularly difficult situations for bounded rationality," as decision-makers search for problems that conform to an established framework (Lee 1993: 143).

[4] Cybernetic learning is change in behavior without the integration of new information, a process that allows for incompatible goals to remain detached and leads organizations to "learn simply what works, without having to understand what works best or even why [their] own behavior is satisfactory" (Lee 1993: 144). This approach to decision-making tries to monitor a small set of variables in order to keep them within acceptable limits. Actors eventually learn to do things unrelated to achieving organizational objectives in order to maintain critical variables.

[5] Single-loop learning occurs when errors that arise in organizations are solved because they were already "anticipated in the underlying theory on which the organization is based." In contrast, double-loop learning requires a critical evaluation of the rules of operation in order to assess potential problems with those theories (Lee 1993: 148).

which involves formulation of a problem, the search for data and alternatives, model building, and decision-making (Quade 1989). While systems analysis is an approach to problem-solving under uncertainty, it has historically left little room for interpretation of uncertain outcomes that cannot be explained by an analyst's models. This is because much of policy analysis continues to concern the anticipatory calculation of outcomes and their comparison to individual (aggregated) intentions (Aldrich 1994). The authors of Chapters 3-5 attempt to address such methodological shortcomings.

The manner in which policymaking is developed through prospective, as opposed to retrospective or on-going analysis, is in part the underlying cause for the limited space for learning available in popular techniques such as operations research, forecasting, and decision-theoretic approaches. Consider a tool of rational choice, the decision tree: all possible courses of action are mapped, the anticipated consequences of possible combinations of choices are summarized, and, given the probabilities of possible outcomes, an "optimal" alternative is selected. Adaptive management has shown that contrary to the assumptions of rational choice, policy learning has to be an on-going, iterative process, which includes a number of characteristics. Learning is based on the daily interactions and routines (i.e., rules, procedures, and technologies) that guide human behavior within a given environment, political institutions derive inferences from history and encode them into routines that guide behavior, learning seeks to address the relation between observed outcomes and group aspirations, and experimentation and the discovery of previously unknown alternatives are used to translate experience into new routines.

While the mechanics of rational choice techniques are predominantly prospective in nature, they also influence the occasions when managers can evaluate past decisions. Crises are assessed by asking such questions as "How can our risk assessment model be better specified to account for unknown concentrations of toxicants?" When these kinds of evaluative questions are raised within policy circles (and often they are not), the expressed purpose is to improve the ability of analysts to continue to operate under previous rubrics of decision-making. The assumptions underlying a prospective, rational approach to policy design (including revealed preferences and the ability to predict policy outcomes), which are grossly violated during periods of policy breakdown such as when a site is contaminated by an industrial process, remain intact to large extent. These assumptions restrict the ability of policymakers to learn from previous decisions or, more importantly, to embed mechanisms for learning into policy design and implementation. Wynne, who has developed a reputation as an expert observer of policy breakdowns, communicates this idea quite effectively when he asserts that crises such as the aftermath of Chernobyl expose the "inability of expert institutions to frankly admit ignorance, contingency and lack of control when appropriate" (Wynne 1996: 13).

Given the lessons learned by adaptive managers, this characterization should come as no surprise. In environments approaching the level of complexity found within the ecosystems embodied by Superfund sites, rational action can be reduced to near unattainability. There are simply too many actors, too many preferences that are either ambiguous or difficult to communicate, and too many unintended consequences for an analyst to proclaim that she has captured the true scope of available policy directions. In response to complexity, it is understandable that policymakers will develop mental models that privilege certain approaches to decision-making while filtering others. This is inevitable to some extent. What should be less readily excused is the manner in which these models discourage learning, not through rational action but through their constraints on the correction of errors and policy change over time. The contributors to *Reclaiming the Land* have recognized the limits posed by traditional tools and methods, and have begun, for example, to rework decision analysis so that it is both systematic and flexible, adaptively manage portfolios of sites, and recast cost-benefit analysis as a consideration of means and ends together in the form of remediation-reuse combinations. They represent a state-of-the-art in terms of the application of decision theoretic approaches to contaminated sites, and afford us an opportunity to assess the extent to which they meet the needs imposed by ecosystem-level management and stewardship.

Chapter 6, "Institutional Controls at Brownfields," broadens the analysis of available methods and tools by focusing on the redevelopment phase of site reuse, drawing on the authors' experience with brownfields. The authors introduce several categories of tools that are more localized than those that comprise Superfund and its focus on cleanup as the ultimate goal for federal involvement with contaminated sites. Redevelopment calls for a set of institutional controls that are not normally associated with CERCLA and RCRA, such as restrictive covenants and easements, permits, consent decrees, and deed notices. These tools are ideally suited for addressing long-term site stewardship, simplifying and streamlining management demands by crafting remedies prospectively to reflect intended reuses as well as the risks typically dealt with by environmental regulators.

Equally important is whether new methods and tools are conducive to the kinds of collaborative processes needed to promote and sustain adaptive management practices at Superfund sites. Can they bridge the many languages and narratives used by different stakeholders, advance common understanding of the techniques of remediation and reuse, and lead to timely error recognition and adjustment? In answering these questions, it is important not to let highly technical approaches crowd out the role of individuals, and non-experts, in decision-making. Institutional change, technical innovation, and deliberative process such as consensus-building discussed in Chapter 7 (which puts into practice Lee's call for negotiation and planning to improve learning in adaptive management) will each benefit from a deeper understanding of the role played by human beings in altering and creating resources, accepting or rejecting modes of conduct, and to various degrees interacting with and shifting the structural constraints imposed from higher levels of the organizations in which they function. Qualitative research methods will be paramount for such a research project. For example, one might pinpoint the agents most responsible for an agency's adaptability, leading to closer examinations of them through ethnographic approaches. And if adaptive management is to ever move toward more cumulative, comparative modes of scholarship, qualitative research will have to be employed to identify important change variables and cases where processes that protect the status quo are met with resistance or even reversed (such as following a crisis at a certain site). Chapter 7, "Rethinking Community Involvement for Superfund Site Reuse," makes extensive use of case study research as well as professional experience at a variety of Superfund sites in order to describe how consensus-building can interact with a series of dynamic variables addressed by site reuse (such as unpredictable timetables and evolving community preferences). Chapter 8, "Toxic Sites as Places of Culture and Memory," adds a unique facet to the participatory dimension of adaptive management – techniques for addressing site histories during the consensus-building process, such as industrial history and historic preservation, which will make it easier to pursue both cleanup and preservation over the long-term.

Bridging diverse approaches and languages with models and methods, questioning the roles and responsibilities that perpetuate conflict, encouraging convergence of different cultures at critical junctures, and developing narratives that encourage a sense of comfort with contaminated sites, preserve traces of the past, and withstand the test of time, are challenges faced by collaborative efforts at Superfund sites. Each of these tasks will only succeed if they are infused with the kinds of institutional support and tools that can encourage group-based decision and learning over time. Our book ends on a hopeful note with a case study of the Tar Creek Superfund site. While Chapter 9, "CHAT: Approaches to Long-Term Planning for the Tar Creek Superfund Site," demonstrates how the Superfund program lacks the ability to adequately address the interaction of contamination in a given medium (soil, water) with human populations (a nonlinear and multigenerational process at megasites such as Tar Creek), a participatory program encouraged by a consortium of researchers, government officials, and local activists is encouraging in its embrace of a more adaptive planning model for cleanup and reuse of the site.

THE ROAD AHEAD FOR SUPERFUND SITE MANAGEMENT

Once a Superfund site is recharacterized as part of a broader ecosystem in which it exists, the central project of adaptive management can be considered and, where appropriate, used by scientists, regulators, and communities. That project is to develop flexible institutions that are attuned to ecosystem dynamics. Superfund sites, existing as part of broader natural and social systems, pose unique challenges to those who seek to manage them and learn from their efforts. As noted, causal understanding emerges at a slower rate than efforts to understand it, and it is not possible for managers to separate the effects of their decisions from those of concurrent shifts in the natural environment.

These and other conditions can create spaces in which reflection and learning occur and allow management processes to take action in light of new information. Experimentation, place sensitivity, multi-scalar analysis, project strategy adjustments, and active participation by relevant stakeholders are essential ingredients to creating such conditions.

We see nascent efforts toward an ecological approach to land use decision-making in zoning that takes natural resources into account (through, for example, a natural resource inventory), watershed management systems, and land use planning that tries to improve site design through such strategies as integrating open space and buffer placement. We even see hints of an understanding in the statute itself that work on a contaminated site should not end with the removal of dangerous substances from water and soil. What remains unavailable is a framework through which long-term site cleanup and stewardship can be pursued.

Over 35 years ago, Caldwell, who advocated the use of ecosystems as a basis for land use policy, asked, "How would a public land policy based upon ecosystems concepts differ from policies based upon other considerations?" (Caldwell 1970: 203). Now, we embark on an effort to ask a similar question of those involved in site remediation and reuse: If we "think like a Superfund site,"[6] and appreciate its nature as dynamic and unfolding at multiple scales of space and time, could we improve the practice of Superfund, encourage learning, and advance a better vision for the places where we live, work, and play? Adaptive management is but a starting point. It supplies us with thirty years' worth of lessons, successes, failures, and questions. Lee suggested over ten years ago that this approach could form the basis for a "civic science," which takes advantage of moments when institutional interests (perpetuated by cognitive biases and cybernetic decision-making) are amenable to change. Crisis, he argued, could afford an opportunity to engage in new forms of

---

[6] Leopold's notion of "thinking like a mountain" was linked to adaptive management and given some technical application by Holling (1978).

learning, promote the use of experimentation, and lower the costs of making mistakes. Now, we have the opportunity to address the next generation of the Superfund program, characterized by long-term site stewardship and reintegration with host communities. *Reclaiming the Land* tries to help present to those at the threshold of this new phase the work that has yet to be done.

REFERENCES

Abbott, A. 1997. "On the Concept of Turning Point." *Comparative Social Research* 16: 85-105.

Aldrich, J. 1994. "Rational Choice Theory and the Study of American Politics." In *The Dynamics of American Politics*, L.C. Dodd and C. Jillson, eds., 208-233. Boulder, CO: Westview.

Ascher, W. 2001. "Coping with Complexity and Organizational Interests in Natural Resource Management." *Ecosystems* 4: 742-757.

Barley, S. 1990. "The Alignment of Technology and Structure Through Roles and Networks." *Administrative Science Quarterly* 35: 61-103.

Bettenhausen, K. and J.K. Murningham. 1985. "The Emergence of Norms in Competitive Decision-making Groups." *Administrative Science Quarterly* 30: 350-372

Bowman, E. and H. Kunreuther. 1988. "Post-Bhopal Behavior at a Chemical Company." *Journal of Management Studies* 25(4): 387-402.

Caldwell, L. 1970. "The Ecosystem as a Criterion for Public Land Policy." *Natural Resources Journal* 10(2): 203.

Carlile, P. 2000. "A Pragmatic View of Knowledge and Boundaries: Boundary Objects in New Product Development." Unpublished manuscript.

Caroll, J. S. 1993. "Out of the Lab and Into the Field: Decision-making in Organizations." In *Social Psychology in Organizations*, K. Murningham, ed., 38-62. Englewood Cliffs, NJ: Prentice Hall.

Denzau, A.D. and D.C. North. 1994. "Shared Mental Models: Ideologies and Institutions." *Kyklos*, 47(1): 3-31.

Environmental Protection Agency. 2002. *Record of Decision: Bunker Hill Mining and Metallurgical Complex OU 3.* http://yosemite.epa.gov/r10/cleanup.nsf /fb6a4e3291f5d28388256d140051048b/cbc45a44fa1ede3988256ce9005623b1!Open Document. Accessed 18 September 2005.

Folke, C. et al. 1998. *The Problem of Fit Between Ecosystems and Institutions.* Working Paper No. 2, International Human Dimensions Program. http://www.uni-bonn.de/IHDP/WP02main.htm (accessed September 20, 2005).

Freedman, P., Nemura, A., and D. Dilks. 2004. "Viewing Total Maximum Daily Loads as a Process, Not a Singular Value: Adaptive Watershed Management." *Journal of Environmental Engineering* 130(6): 695-702.

Friedman, J. 1987. *Planning in the Public Domain: From Knowledge to Action.* Princeton: Princeton University Press.

Friedman, R. 1993. "Bringing Mutual Gains Bargaining to Labor Negotiations: The Role of Trust, Understanding, and Control." *Human Resources Management* 32(4): 435-459.

Galison, P. 1979. "Trading Zone: Coordinating Action and Belief." In *The Science Studies Reader*, M. Biagioli, ed., 137-160. New York: Routledge.

Gersick, C. J. and J. R. Hackman. 1990. "Habitual Routines in Task-Performing Groups." *Organizational Behavior and Human Decision Processes* 47: 65-97.

Gulliver, P. 1979. *Disputes and Negotiation: A Cross-Cultural Perspective*. New York: Academic Press.

Hoffman, R. L. & N. R. Maier. 1979. "Valence in the Adoption of Solutions by Problem Solving Groups: Concepts, Methods, and Results." In *The Group Problem-Solving Process*, L. R. Hoffman, ed., 17-30. New York: Praeger.

Holling, C. S. 1978. *Adaptive Environmental Assessment and Management*. New York: Wiley.

Holling, C. 1992. "Cross-Scale Morphology, Geometry and Dynamics of Ecosystems." *Ecological Monographs* 62 (4): 447-502.

Holling, C. 1995. "What Barriers? What Bridges?" In *Barriers and Bridges to Renewal of Ecosystems and Institutions*, H. Gunderson, C. Holling, and S. Light, eds., 3-34. New York: Columbia University Press.

Holling, C. 1996. "Engineering Resilience Versus Ecological Resilience." In *Engineering Within Ecological Constraints*, P. C. Schulze, ed. Washington, DC: The National Academy Press.

Holling, C. and G. Meffe. 1996. "Command and Control and the Pathology of Natural Resource Management." *Conservation Biology* 10: 328-337.

Imperial, M. 1999a. "Institutional Analysis and Ecosystem-Based Management." *Environmental Management* 24: 449-465.

Imperial, M. 1999b. "Analyzing Institutional Arrangements for Ecosystem-Based Management: Lessons from the Rhode Island Salt Ponds SAM Plan." *Coastal Management* 27: 31-56.

Johnson, K. et al. 1999. *Bioregional Assessments: Science at the Crossroads of Management and Policy*. Washington, DC: Island Press.

Landy, M., M. Roberts, and S. Thomas. 1994. *The Environmental Protection Agency: Asking the Wrong Questions from Nixon to Clinton*. New York: Oxford University Press.

Lee, K. 1991. "Rebuilding Confidence: Salmon, Science, and Law in the Columbia Basin." *Environmental Law* 21: 745.

Lee, K. 1993. *Compass and Gyroscope: Integrating Science and Politics for the Environment*. Washington, DC: Island Press.

Lee, K. 1999. "Appraising Adaptive Management." *Conservation Ecology* 3(2): 3.

Lee, K. and J. Lawrence. 1986. "Adaptive Management: Learning from the Columbia River Basin Fish and Wildlife Program." *Environmental Law* 16: 431-460.

Lessard, G. 1998. "An Adaptive Approach to Planning and Decision-making." *Landscape and Urban Planning* 40: 81.

Levitt, B. and J. March. 1988. "Organizational Learning." *Annual Review of Sociology* 14: 319-340.

Light, S. 2002. "Adaptive Management: A Valuable but Neglected Strategy." *Environment* June 2002: 42-43.

March, J. and J. Olsen. 1989. *Rediscovering Institutions*. The Free Press.

March, J. and H. Simon. 1958. *Organizations*. New York: Wiley.

McDaniels, T. and R. Gregory. 2004. Learning as an Objective Within a Structured Risk Management Decision Process. *Environmental Science and Technology* 38(7): 1921.

McGrath, J. 1984. *Groups: Interaction and Performance*. Prentice Hall.

Meyer, J. and B. Rowan. 1977. "Institutionalized Organizations: Formal Structure as Myth and Ceremony." *American Journal of Sociology*, 83: 340-363.

Meyerson, D., K. Weick, and R. Kramer. 1996. "Swift Trust and Temporary Groups." In *Trust in Organizations*, R. Kramer & T. Tyler, eds. Thousand Oaks: Sage.

Moir, W. and W. Block. 2001. "Adaptive Management on Public Lands in the United States: Commitment or Rhetoric?" *Environmental Management* 28(2): 141-148.

National Research Council. 1994. *Rangeland Health: New Methods to Classify, Inventory, and Monitor Rangelands*. Washington, DC: National Academy Press.

National Research Council. 2005. *Superfund and Mining Megasites – Lessons from the Coeur d'Alene River Basin*. Washington, DC: National Academy Press.

Norton, B. and A. Steinemann. 2001. "Environmental Values and Adaptive Management." *Environmental Values* 10 (2001): 473-506.

Oliver, C. 1991. "Strategic Responses to Institutional Processes." *Academy of Management Review* 16(1): 145-179.

Ostrom, E. 1990. *Governing the Commons: The Evolution of Institutions for Collective Action*. Cambridge University Press.

Pauchant, T.C. and I. Mitroff. 1992. *Transforming the Crisis-Prone Organization: Preventing Individual, Organizational, and Environmental Tragedies*. San Francisco: Jossey-Bass Publishers.

Phillips, C. and J. Randolph. 2000. "The Relationship of Ecosystem Management to NEPA and its Goals." *Environmental Management* 26(1): 1-12.

Prichard, L., C. Folke, and L. Gunderson. 2000. "Valuation of Ecosystem Services in Institutional Context." *Ecosystems* 3: 36-40.

Quade, E. 1989. *Analysis for Public Decisions*. Englewood Cliffs, NJ: Prentice-Hall.

Shepherd, A. and C. Bowler. 1997. "Beyond the Requirements: Improving Public Participation in EIA." *Journal of Environmental Planning and Management* 40(6): 725-738.

Shrivastava, P. 1988. "Understanding Industrial Crises." *Journal of Management Studies* 25(4): 283-303.

Staw, B.M., L.E. Sandelands, and J.E. Dutton. 1981. "Threat-Rigidity Effects in Organizational Behavior: A Multilevel Analysis." *Administrative Science Quarterly* 26: 501-524.

Steyer, G. and D. Llewellyn. 2000. "Coastal Wetlands Planning, Protection, and Restoration Act: A Programmatic Application of Adaptive Management." *Ecological Engineering* 15: 385-395.

Tarrow, S. 1990. "Mentalities, Political Cultures, and Collective Action Frames: Constructing Meanings Through Action." In *Frontiers in Social Movement Theory*, A. Morris and C. Mueller, eds., 174-202. New Haven: Yale University Press.

Torrell, E. 2000. "Adaptation and Learning in Coastal Management: The Experience of Five East African Initiatives." *Coastal Management* 28: 353-363.

Walters, C. 1986. *Adaptive Management of Renewable Resources*. New York: Macmillan.

Walters, C. 1997. "Challenges in Adaptive Management of Riparian and Coastal Ecosystems." *Conservation Ecology* 1(2): 1.

Walton, R. and R. McKersie. 1965. *A Behavioral Theory of Labor Negotiations*. Beverly Hills: Sage Publications.

Weick, K. 1988. "Enacted Sensemaking in Crisis Situations." *Journal of Management Studies* 25(4): 305-317.

Weick, K. 1993. "The Collapse of Sensemaking in Organizations." *Administrative Science Quarterly* 38: 628-652.

Wicks, D. 2001. "Institutionalized Mindsets of Invulnerability: Differentiated Institutional Fields and the Antecedents of Organizational Crisis." *Organization Studies* 22(4): 659-692.

Wilhere, G. 2002. "Adaptive Management in Habitat Conservation Plans." *Conservation Biology* 16(1): 20-29.

Wynne, B. 1996. "Patronizing the Public." *Times Higher Education Supplement*, Apr. 12: 13.

Zucker, L. 1987. "Institutional Theories of Organization." *Annual Review of Sociology* 13: 443-464.

1

OVERVIEW OF THE SUPERFUND PROGRAM

Alexander E. Farrell
University of California at Berkeley Energy and Resources Group

INTRODUCTION

Beginning in the nineteenth century, modern science and industry introduced compounds into the environment not found in nature but useful for their new properties, such as persistence and ability to control pests. Unfortunately, these same properties make these materials potential hazardous contaminants. As industrial processes in the United States grew in size and began to use greater amounts of hazardous substances, contemporary waste management practices, described as "cheap and casual" by Andrews (1999: 245), were applied to hazardous materials as well. However, the effects of hazardous substances could be very different from those of traditional wastes, for which odor and infectious disease were the principal problems, so these practices resulted in significant potential health and environmental risks (Hays 1987, Ch. 6). Up through the 1960s, the lack of awareness of the potential risks of hazardous wastes resulted in many abandoned hazardous waste sites (Hird 1994).

Growing public awareness that areas such as the Love Canal neighborhood in New York, the "Valley of the Drums" in Kentucky, the Stringfellow Acid Pits in California, and other sites across the United States were contaminated with hazardous substances, much of it industrial waste, sparked a national controversy. Dramatic events, like the 1978 fire at an illegal hazardous waste dump in Chester, Pennsylvania, that hospitalized over forty firefighters, only added to the sense of urgency (Wildavsky 1995). The ensuing debate over how best to deal with these problems led to the creation of the Superfund program under the Comprehensive Environmental Response, Compensation, and Liability Act (CERCLA) in 1980, and the Superfund Amendments and Reauthorization Act (SARA) in 1986. Together, these and related laws established a federal program for preventing, mitigating, and responding to releases of hazardous substances that might threaten human health and the environment.

Contamination with hazardous substances is a massive problem. Over the last 24 years, the Superfund program has responded thousands of times to releases (or potential releases) of hazardous substances into the environment and continues to respond to over 300 new (or newly discovered) releases every year. These actions have reduced, halted, or prevented the exposure of millions of people to hazardous substances and permanently destroyed or isolated many thousand tons of hazardous materials. R. N. Andrews referred to CERCLA, SARA, and related laws when he noted that "the transformation of waste management practices was one of the most impressive yet least noted successes of American environmental policy" (Andrews 1999: 249).

The most expensive and controversial part of the Superfund program is the National Priorities List (NPL) of sites and the two mechanisms for cleaning up these sites: liability rules that enable the EPA to sue polluters in order to recover the costs of cleaning up NPL sites, and a federal trust fund to pay for cleanups for which no potentially responsible party (PRP) can be found.

While a great deal has been written about the NPL program, the other aspects of the Superfund program are poorly understood (exceptions include Wildavsky 1995: 153, 186; Probst and Konisky 2001: Ch. 3; Anderson, Thompson, and Suk 2002). Indeed, the term "Superfund site" is often used to mean "NPL site," although, as is discussed below, the vast majority of sites at which the Superfund program participates in a cleanup of some sort are *not* NPL sites. There is some justification for this focus on the NPL; it is by far the most expensive part of Superfund, and the liability provisions are extremely stringent – CERCLA creates strict, joint, and several liability for PRPs.

However, the Superfund program also includes a number of other, less well known components, and these components may offer more flexibility and adaptability in managing hazardous substances than is often assumed. These other components also provide many additional capabilities. For instance, the Superfund program supports communities that are burdened with the presence of hazardous material sites so they can better understand and participate in decisions about what to do with them. Superfund created a program for developing and deploying knowledge and technologies to better manage hazardous substances. It provided training for thousands of first responders (firefighters, police, emergency room nurses, etc.) so they could detect and identify hazardous substances in order to protect themselves and the public. It has enabled the restoration of hundreds of communities and ecosystems. Finally, Superfund created a powerful incentive for innovation to reduce the need for hazardous substances in the economy and the amount of hazardous waste that is generated.

The Superfund program has led to cleanup or other remediation at many sites across the country, which has reduced human health risks for cancer, lead poisoning, acute injuries involving hazardous substances, and probably birth defects. In addition, these actions have improved environmental quality at many sites and protected a great deal of the nation's groundwater. CERCLA, SARA, and related laws have also increased knowledge about and capability to deal with accidents involving hazardous substances through research, development, and training. Recently, these capabilities have proven useful in counter-terrorism planning and response.

This chapter will sketch the current Superfund program, defined as the entire system of regulations and activities that have come into being as a result of the CERCLA and SARA statutes. This will serve as an introduction to the key concepts, terms, and controversies that are evaluated in detail in the succeeding chapters. Most of the other chapters in this book deal with "Superfund sites" as typically defined, that is, they deal with NPL sites. However, the concepts of adaptive management they develop may be applicable to other, less problematic "brownfield" sites as well. In this way, the other chapters in this book provide broad, generally applicable ways of thinking about how to improve the management of hazardous substances at many types of sites.

THE NEED FOR SUPERFUND

The Superfund program addresses the problem of actual or potential uncontrolled releases of hazardous substances into the environment.[1] By the time CERCLA was passed in 1980, improvement of hazardous waste management in the United States was already under way, following the passage of the Toxic Substances Control Act (TSCA) and the Resource Conservation and Recovery Act (RCRA) in 1976. These laws governed the active production and controlled release (such as landfill disposal) of hazardous substances. However, there was growing evidence that substantial quantities of uncontrolled hazardous substances existed in places and in conditions throughout the United States that could present human health and ecological risks, or might reasonably be expected to do so in the future. These

---

[1] Although it is common to use the terms "hazardous substance," "hazardous material," and "hazardous waste" interchangeably, these terms have different statutory definitions. CERCLA and SARA authorize EPA to address hazardous *substances*, including wastes as well as other types of substances (e.g., product spills), but *excluding* petroleum and petroleum products. Oil spills are dealt with under the Oil Pollution Act by agencies authorized to address hazardous *materials*. The management of hazardous *wastes*, including the treatment, storage, and disposal of hazardous wastes, is regulated by the Resource Conservation and Recovery Act. CERCLA and SARA deal with *uncontrolled* releases of hazardous substances, both wastes and non-wastes.

hazardous substances were the result of *prior* actions, which neither TSCA nor RCRA addressed directly. Moreover, it was clear that many sites with hazardous substances at them had *potential* uncontrolled releases that had not yet leaked or spilled. Lastly, it was clear that accidents and illegal activities also created new uncontrolled releases (Landy, Roberts, and Thomas 1994: Ch. 5).

It is important to place the concepts of *toxic, hazard,* and *risk* into perspective (Paustenbach 2002). *Toxicity* is an inherent property of all substances; that is, any chemical can cause adverse effects in sufficient concentrations (i.e., "the dose makes the poison"). In contrast, *hazards* are specific situations that raise the likelihood or severity of an adverse outcome, such as exposure to a substance at concentrations that could lead to harmful effects. The term *risk* is used to refer to the probability that an adverse health outcome will occur in a person or group exposed to a specific concentration of a hazardous agent, or that adverse ecological effects will occur.

The principal inherent dangers presented by hazardous substances are negative health effects, including both acute effects (e.g., acute poisoning, injuries from fires or explosions) and a variety of long-term effects (e.g., cancers, birth defects). (See Johnson 1999; Bove, Shim, and Zeitz 2002; Dolk and Vrijheid 2003.) Hazardous substances found at Superfund sites that cause such effects include: lead, arsenic, benzene, trichloroethylene, mercury, and over 250 other hazardous substances (Agency for Toxic Substances and Disease Registry 2003a, 2003b).

In order for a hazardous substance to present a health risk, a completed exposure pathway for exposure to that substance must exist. Since 1990, completed exposure pathways for hazardous substances have been found at over 15,000 sites (NPL and non-NPL) in the United States (Agency for Toxic Substances and Disease Registry 2003a, 2003b). Exposure to hazardous substances varies significantly from site to site, and human exposure to hazardous substances may occur through multiple routes. Data on human exposure due to uncontrolled releases of hazardous substances exist for some cases, but the type of exposure data needed for analysis of expected risk are not collected.[2] Overall, the amount of data on exposures to hazardous substances is extremely poor and frustrates research efforts that attempt to quantify it (Harrison 2003). However, a lack of exposure and risk data does

---

[2] Specifically, exposure and risk information for the maximally exposed individual (MEI) exists for most sites on the National Priorities List (NPL), but these data are contained in individual baseline risk assessments for each site and are not compiled in a single place, so are not readily accessible. Further, neither data for typical individuals nor population exposure data exist for these sites, and even less information is available for non-NPL sites with uncontrolled releases of hazardous substances, which are far greater in number.

not imply there are no exposures or risks – the Superfund program is designed to identify actual or potential exposures and then remedy those situations; it is *not* designed to collect detailed exposure information. Further, it is not clear that there exist ethical methods that could be used to collect exposure data – doing so might require allowing exposures to knowingly continue, thus placing the exposed persons at higher risk. Sometimes it may be possible to reconstruct exposures after the fact, but where this is not possible, the only alternative might be to allow exposed populations to remain exposed for some time while the data are collected, which is generally considered unethical.

Uncontrolled releases of hazardous substances can also damage ecological systems that provide services to both humans and other species. Examples of ecological risks include contamination of ground water, wetlands, lakes and rivers, estuaries, and grasslands (Jones et al. 1999; Morey et al. 2002). This contamination can reduce organism survival and growth rates, change species composition, reduce ecosystem productivity, and have other effects which can lead to reductions in valued ecosystem services such as water filtration, nutrient cycling, fishing, and use of habitat.

But why does the government need to become involved at all in this issue? The answer flows directly from standard economic theory. Exposures to hazardous substances generally fall into the category that economists call *externalities,* or effects that are created by economic activity (e.g., manufacturing) but are not included in the decision-making about or the cost assigned to that activity. Superfund was designed to implement remedies to cases where the activities had taken place in the past, creating abandoned sites with actual or potential releases of hazardous substances, an obvious externality. Government action is usually required to correct externalities (Baumol and Oates 1988).

An additional problem is the lack of incentives for discovery and innovation that are aimed at providing public goods, such as environmental quality (Orr 1976; Baumol and Oates 1988; Jung, Krutilla, and Boyd 1996). Public goods are products or services that if supplied to one person are available to others at no extra cost. Generally, public goods are considered non-rival in that consumption by one person does not reduce the amount available to others, and are considered non-excludable in that the producer is unable to prevent anyone from consuming it. For these reasons, markets in public goods rarely exist and there are few incentives for discovery and innovation that are aimed at producing the public goods. Similarly, markets for environmental information generally do not exist either. Thus, private firms tend to find it uneconomic to invest in research and development efforts to develop environmental information and technologies. Those that improve environmental quality tend to require government action (Hamilton 1995; Konar and Cohen 1997; Fung and O'Rourke 2000; Taylor, Rubin, and

Hounshell 2003). This leads to a need for government action in order to develop technologies that provide public goods.

For sites contaminated with hazardous substances, the cost of remedial action is generally not justified by increased returns in real estate markets. It is ludicrous to imagine that, without Superfund, manufacturing companies, property developers, mortgage holders, and other firms would have responded to the problem of contamination with hazardous substances by investing in the research and development efforts that have been needed over the last several decades to deal with this problem. Similarly, it is not clear what organization would have created the emergency response capabilities that Superfund has helped come into being – municipal fire departments and medical response teams would have been on their own to deal with hazardous substances. And a world without Superfund would leave no private incentives for the remediation of releases of hazardous substances into ecosystems. Thus, government action is required both to learn about contaminated properties and to develop and deploy the technologies needed to remedy the problems.

Several different approaches are available to government: it can act directly to correct the externality; it can create standards or procedures for private actors (individuals and firms) to correct the externality; or it can impose requirements on the free market that internalize the external costs. The Superfund program uses all three approaches. For some sites, the EPA conducts waste isolation or cleanup, both at NPL and other sites (e.g., sites managed through a removals program). In addition, CERCLA and SARA authorize research, development, and training programs. Enforcement activities authorized by these statutes embody the second approach, whereby the EPA or other agencies (notably the National Oceanographic and Atmospheric Agency and the Department of the Interior) sue polluters in order to get them to remediate releases. The liability provisions of CERCLA are powerful mechanisms for internalizing the otherwise external costs of potential releases into decisions about the management of hazardous substances in private transactions.

Views about the Superfund program and appropriate approaches to managing hazardous substances are often dependent on whether one focuses on hazards or risks. Some observers seem to feel that the mere existence of hazardous substances in environmental media is problematic and requires a remedy. According to this view, it seems, any potential for the exposure of human or ecological receptors to hazardous substances represents an unacceptable legacy to future generations. For people who hold these views, such concerns about hazards typically override concerns about the cost of remedies.

In contrast, other observers focus on reducing risks and seem to feel that a more technically oriented approach that measures and balances risks and costs is most appropriate. The presence of hazardous substances alone is insufficient to warrant action and the presence of exposed receptors and risk is needed. However, according to this view, even an elevated risk may be insufficient (because such risks need to rise above some threshold, such as a social norm for risks that require government action) to justify the cost of a remedy.

APPROACHES TO MANAGING HAZARDOUS SUBSTANCES

CERCLA and the various Superfund programs address the problem of uncontrolled releases of hazardous substances using six different approaches that have fairly different characteristics. The federal government implements many of these approaches, but states and private firms also play key roles. CERCLA and SARA are the authority for all of the federal actions, and they provide strong support for many state and private actions. The federal government also provides significant budgetary support for state environmental programs. Hence, the state and private actions can be partially attributed to the federal statutes. Together, federal and state governments and private industries take actions under Superfund, which can be classified into six basic approaches, as defined in Table 1.1.

Response is the most direct and obvious of the approaches taken under the Superfund program, and by far the most expensive, accounting for perhaps as much as 90 percent of all resources (public and private) expended under CERCLA and SARA. Superfund is not really a regulatory approach, but encompasses a broad set of activities that can be grouped into two sets, based largely on size and underlying legal authority. Superfund responses are designed to address the continuum of health and environmental risks ranging from emergencies to long-term problems. Technical options include containment, chemical neutralization, biodegradation, incineration, groundwater treatment, institutional controls (e.g., temporary access control by fencing or permanent restrictions on activities such as digging), and others. Statutory authority for removal actions, and in particular time-critical removals, provides for rapid response where the problem is urgent. Because response actions are the most complex and expensive part of the Superfund program, they are described in detail in the next section of this introductory chapter.

The second approach includes efforts to improve the involvement of communities near remedial action sites. These efforts help individuals, families, and communities understand sites near them, and participate in Agency decisions about those sites.

The third approach includes enforcement efforts, which have brought far more resources to bear on the problem of releases than the federal government had available. This has led to many more response actions and the cleanup of many more contaminated sites. In addition, enforcement activities help ensure that the parties responsible for the contamination problem pay the costs of cleaning it up. That is, enforcement helps make certain that, as much as possible, the "polluters pay." Much of the authority for enforcement derives from the stringent liability provisions of Superfund, along with the enforcement provisions. These provisions are also powerful incentives for private innovation in products and processes that need fewer hazardous substances as inputs and produce less hazardous waste, although RCRA and other laws also contribute to this effect. These provisions also serve as a backstop to state response programs and help encourage private firms to respond to releases on their own.

The fourth approach to addressing the problem of uncontrolled releases of hazardous substances is research and development, which are conducted by EPA's Office of Research and Development (ORD) and Environmental Response Team (ERT), the Agency for Toxic Substances and Disease Registry (ATSDR), and the National Institutes of Environmental Health Sciences through the Superfund Basic Research Program (SBRP). These organizations also engage the fifth approach, training for a variety of groups, including first responders (e.g., firefighters), first receivers (i.e., emergency room staff), and scientists. While there are still limitations in the understanding of hazardous material risks and in the methods and technologies for managing releases, these research and training efforts have gone a long way toward improving our scientific knowledge and practical capabilities since 1980.

The fifth approach, called training, involves efforts to make professionals in many different fields more capable of identifying and responding safely to uncontrolled releases of hazardous substances. It differs from the empowerment approach in that it is focused on professionals and relevant organizations like hospitals and municipal governments, not the public. Many different Superfund-supported organizations conduct training in this sense, including especially the Environmental Response Team (ERT) and ATSDR. These activities include training to deal with some types of homeland security issues, for instance, attacks with biological agents.

The sixth approach is natural resource restoration,[3] which frequently occurs at NPL sites, but which can also occur at accidental spills. These efforts are aimed mainly at restoring ecological functions that have been damaged or destroyed by uncontrolled releases of hazardous materials.

---

[3] Executive Order 12316 delegated the Presidential authorities of CERCLA to various federal agencies. While EPA is charged with implementing most of the response provisions of CERCLA and many of the enforcement provisions, the natural resource damages provisions of trustees are assigned to the various federal agencies (e.g., the Departments of Agriculture, Commerce, and Interior).

Natural resource restoration has become an increasingly important approach taken under Superfund in the last decade. Natural resource restoration activities are undertaken by organizations that act as public trustees, including several federal departments (e.g., Interior), states, and tribes, but not EPA.

Table 1.1. Superfund approaches.

Name	Description
Response	Remedial Activities: Activities associated with sites (including the NPL, non-NPL federal sites, state sites, private sites) and spills to reduce the amount, toxicity, and mobility of hazardous substances in order to reduce human health and ecological risks. As used here, "cleanup" includes institutional controls designed to prevent exposure. Remedial actions tend to address only actual releases, but may address potential releases as well.
	Removals: Activities designed to address immediate human health risks due to uncontrolled releases of hazardous substances. Removals may occur at all types of sites and spills, and may be taken by federal or state agencies. Emergency responses include actions taken following terrorist attacks. Removal actions may address either potential or actual releases, and account for most of the potential releases addressed by response actions.
Community Involvement	Activities that help citizens and businesses located near sites undergoing cleanup to better understand and participate in the process.
Enforcement	Actions taken by federal and state governments to effect response actions by potentially responsible parties, to recover costs of federal and state responses, and to restore natural resources.
Research and Development	Efforts to better understand hazardous substances and their effects on human health and the environment, to develop new technologies and strategies for reducing the risks of hazardous substances, and to lower the cost of cleanup.
Training	Activities designed to improve the capability of professionals (e.g., paramedics and firefighters) and organizations (e.g., hospitals and municipal governments) that may be required to address releases of hazardous substances, often state and local first responders and first receivers (i.e., hospital emergency departments). Includes homeland security preparedness.
Natural Resource Restoration	Actions taken to return ecological features (e.g., rivers, prairie, scenic vistas) back to conditions similar to those before hazardous substances were introduced and thus to restore the flow of valued services (e.g., fishing, tribal uses, wildlife habitat, protection of resources for future generations).

Response Actions

Of the approaches used in the Superfund program described above, responses, including both removal actions and remedial actions, are the most complex, accounting for a majority of Superfund spending. Superfund responses address the continuum of health and environmental risks ranging from emergencies to long-term problems. This section describes Superfund response actions.

The National Oil and Hazardous Substances Pollution Contingency Plan (NCP, 40 CFR 300) is the regulation that specifies how CERCLA remedial and removal actions are conducted. The NCP was first established in 1968 to deal with oil spills. CERCLA added authority to respond to uncontrolled releases of hazardous substances in a manner consistent with the NCP. Today the U.S. Coast Guard and the Environmental Protection Agency (EPA) manage the NCP jointly, and the program involves over 16 federal agencies, as well as many state and local representatives. The NCP provides a national framework for emergency response capabilities and promotes coordination among the hierarchy of responders and contingency plans. A Superfund response is triggered by the discovery of a hazardous substance release, or a substantial threat of a release. CERCLA Section 101 defines a release as "any spilling, leaking, pumping, pouring, emitting, emptying, discharging, injecting, escaping, leaching, dumping, or disposing into the environment (including the abandonment or discarding of barrels, containers, and other closed receptacles containing any hazardous substance or pollutant or contaminant)." The release must be of a hazardous substance as defined in CERCLA or must present an imminent or substantial danger to public health or welfare. Petroleum spills are specifically excluded from the Superfund.

Based on these regulations and complementary state laws and programs, EPA and the states have over time crafted a set of response options that are flexible and that maximize PRP involvement in response actions. In situations where there is an obvious, immediate health risk, the NCP authorizes limited federal expenditures to deal with the problem. In situations where the cost of remediating a site is larger, the NCP requires more testing and analysis to determine more definitively the nature and extent of contamination and select the best method for dealing with the release. In both cases, the liability provisions of CERCLA and SARA apply, so PRPs can be required to reimburse EPA for the cost of remediation. This highlights the importance of the federal trust fund; it can be used to begin to address without delay releases that could be harmful to the public, and then the costs incurred can be recovered from the PRPs. In part to avoid the liability provisions of the Superfund program and in part to restore more power to local decision-makers, the states have developed complementary programs to deal with releases, and PRPs have undertaken voluntary remedial actions (usually under the supervision of state Voluntary Cleanup Programs, or VCPs).

Response actions are divided into two types: removal actions and remedial actions.

Removal actions, in particular time-critical removals, provide for rapid response where the problem needs to be addressed in an urgent manner. Typically, removal actions respond to chemical spills, human health threats that might cause harm from short-term exposures (e.g., lead-contaminated residential soils), or situations that may cause a sudden release (e.g., leaking drums). Due to the necessity for quick action, removals are usually conducted without significant administrative or planning activities and relatively little documentation.

Remedial actions (and some removal actions, called non-time-critical removals) address situations where the response can be taken in a more deliberate fashion, allowing for more in-depth planning, evaluation, and documentation. Remedial actions are limited by regulation to sites on the NPL.

However, the distinction between remedial actions and removals is not as sharp as it might seem. The types of response actions that can be taken (e.g., waste treatment, excavation and disposal, providing alternate water supplies) are identical under both sets of authorities, except that permanent relocation of residents is only specifically authorized as a remedial action. In practice, the removal program is often used to address completed exposure pathways with higher levels of exposure, while the remedial program addresses risks where there are no current exposures or where the levels of exposure allow for a more deliberate planning process.

A simplified diagram of the processes (or "pipelines") for the various Superfund responses is shown in Figure 1.1.[4] Note that there are three essentially separate pipelines: a site screening or pre-remedial action (left pipeline); a removal action (upper right pipeline); and the remedial action (lower right pipeline). The site screening pipeline is used to sort out the many notifications and discoveries that are referred to EPA to ensure that each site receives an appropriate response, or receives no response if that is appropriate. It is important to keep in mind while reading the description of the site screening pipeline (as well as the NPL pipeline) that at any time during these processes the need for a removal action may arise, in which case the screening and analysis are temporarily stopped to deal with whatever immediate health risk has created the need for the removal action.

---

[4] More information can be found at
www.epa.gov/superfund/action/process/sfprocess.htm.

Figure 1.1. Superfund response pipelines.

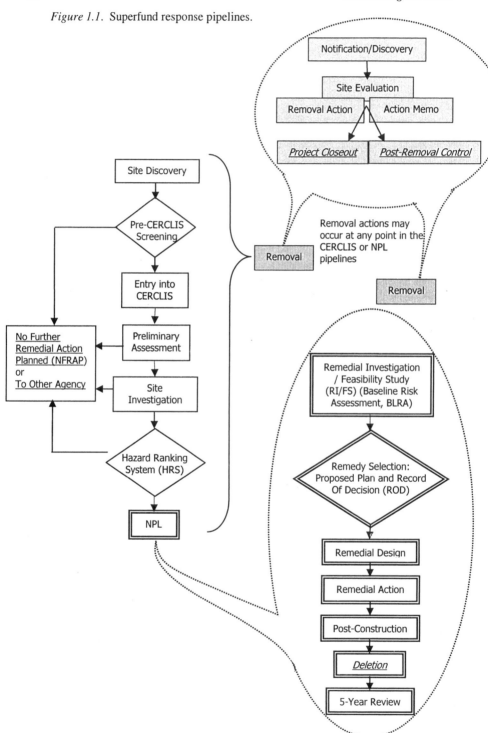

The use of the site screening process through the NCP has helped reduce the uncertainty associated with actual and potential releases. Local first responders (e.g., firefighters, police) and others who discover what they believe to be a hazardous substance release have a single place to report the discovery: the National Response Center (800-424-8802 or www.nrc.uscg.mil/nrchp.html). As a result, adequately trained personnel determine if the substance is hazardous, and if so, what to do about it. Many sites that were suspected of contamination by hazardous waste have been assessed through this mechanism, and most of them were found not to qualify for the NPL. Thus, the site screening process reduces the uncertainty associated with many potentially contaminated sites.

Removal Actions

The NCP gives EPA the authority to respond rapidly to urgent problems associated with releases, such as chemical spills, human health threats that might cause harm from short-term exposures (e.g., lead-contaminated residential soils), and situations that may cause a sudden release (e.g., leaking drums). A wide variety of response actions can be taken under this authority, such as waste treatment, excavation and disposal, or providing alternate water supplies. However, less permanent measures, such as erecting a fence to prevent access to the contaminated area and thus prevent exposure, are also common removal actions. In part, this is due to the limited monetary and temporal scope permitted for removal actions ($2 million and one year, with occasional exceptions), and in part due to the way EPA coordinates removal actions with larger, more complex remedial actions to ensure health and ecological risks are dealt with appropriately at all sites.

As a result of time and spending limits, removal actions do not support the detailed investigation and planning needed to ensure that the locations of all hazardous substances on a site are identified, that the extent of contamination is fully characterized, or that all the hazardous substances on the site are treated. For this reason, detailed risk information is generally not available for releases associated with removal actions. In practice, the removal program is often used to address completed exposure pathways with high levels of exposure, and the remedial program is used to address such risks where there are future risks but no (or limited) current exposures, or where a completed exposure pathway is interrupted temporarily (e.g., with a fence).

In some cases (slightly less than 20% of the time) PRPs undertake removal actions with EPA or state supervision (Probst and Konisky 2001: 25).[5] In these cases, the limits on expenditures and duration do not apply.

EPA classifies removal actions into two types: time-critical, and non-time-critical. To provide an example of typical removal actions, parts of a table from Probst and Konisky (2001: 20-21) are reproduced below in Table 1.2.

Table 1.2. Examples of removal actions.

Type of Removal	Example Actions
Time-Critical	<ul><li>Respond to truck accidents and train derailments involving chemical releases</li><li>Temporarily provide bottled water to homes with contaminated water supplies</li><li>Remove and dispose of chemicals abandoned by roadsides or in vehicles</li><li>Respond to fires and explosions involving chemicals at operating or abandoned facilities, tire fires, and so forth</li><li>Clean up and monitor mercury contamination at schools and private residences where children have played with metallic mercury</li><li>Clean up and monitor chemical releases due to natural disasters (e.g., floods)</li><li>Restrict access to and remove and dispose of chemicals at abandoned or bankrupt facilities or warehouses that may be subject to vandalism or fires (e.g., small electroplating shops, illegal drug production sites)</li><li>Stabilize mining wastes to prevent releases to surface and groundwaters</li></ul>
Non-Time-Critical	<ul><li>Remove "hot spots"</li><li>Install groundwater treatment systems for contaminated groundwater in conjunction with the remedial program</li></ul>

Source: Probst and Konisky 2001, Table 2-3

As of the end of FY 2003, EPA had completed approximately 7,400 removal actions.[6] Over the last decade, an average of 430 removals has

---

[5] Note that Probst and Konisky excluded federal facilities from most of their analyses, which may lead to differences in total cases they report versus those of other researchers.

[6] Data obtained from EPA's internal CERCLIS database on February 4, 2004. A small number (< 3%) of the sites in this database are double-counted due to

occurred annually. Based on records in CERCLIS, 69% of removal actions were time-critical through the end of 2002. Removal actions are often responses to spills and other accidents (which are not sites in the usual sense), so first responders discover many of these releases; others are reported by facility personnel to comply with statutory and NCP requirements. Other removal actions are taken at sites that EPA expects to eventually undergo remedial action on the NPL, but which could become significantly worse before the remedial action process reaches the action phase. Removal actions can take place at NPL sites (or sites that will eventually be placed on the NPL) at any time after discovery.

Remedial Actions

The NCP gives EPA the authority to undertake larger, more complex responses to actual or potential uncontrolled releases of hazardous substances through the remedial action program.

Remedial actions also tend to fulfill the Congressional mandate in SARA to seek permanent solutions to releases at sites where both removal and remedial actions occur. This can be a major distinction. The removal action(s) may temporarily interrupt an exposure pathway, but leave the contamination in place or only remove some of the contamination. The remedial action then treats or isolates the hazardous substance, thereby providing a long-term solution. Often, soil or groundwater contamination is dealt with through remedial actions because these pathways may not present an imminent risk (which would trigger a removal action) or are expensive and time-consuming to address (thus falling outside of the limitations of removal actions). Remedial actions sometimes involve isolating hazardous substances on-site in specially engineered systems. In this sense, it is not completely accurate to say that all Superfund sites are "cleaned up"; in some cases the hazardous materials are simply isolated from contact with people and the environment. The permanence of these engineered systems is reviewed on a regular basis (every 5 years) after the site work is completed.

Sites must be on the NPL for EPA to have the authority to conduct a remedial action and for the liability scheme to be invoked against potentially responsible parties. Thus, NPL sites are the best known and most expensive part of the Superfund program, and are often what is being referred to when terms like "Superfund cleanups" and "Superfund sites" are used casually. In recent years, more than two-thirds of all remedial actions have been paid for by PRPs, who also pay for remedial action at VCP sites (Probst et al. 1995: 33). The cost of individual NPL remedial actions varies a great deal, but

changing designations. The value above is 97% of the values reported in the database.

estimates of average costs for individual sites (excluding overhead) fall in a fairly narrow range of about $15-$30 million, with a best subjective estimate of about $25 million; this is around 25 times the cost of a removal (Probst et al. 1995: 33; Hamilton and Viscusi 1999: 111, 119). However, some individual remedial actions involve hundreds of millions of dollars of effort, so adequate preparation and design is often a complex task in itself.

The NPL was originally created as a list of the worst hazardous substance sites in the country, but in the last decade many of the most serious problems have come to be addressed by state programs, often overseeing private remedial actions, and the NPL has become a tool for addressing the subset of worst sites at which federal resources are needed (e.g., abandoned sites), or at which federal enforcement powers are needed.

Thus, the primary distinction between NPL sites and other hazardous waste sites is often a question of whether federal authorities or resources will be needed, not the risks presented by the site, or even the cost of remedial action. For this reason, while NPL sites tend to have serious contamination problems, not all sites with serious contamination end up on the NPL, and the NPL does not include all of "the worst of the worst" (General Accounting Office 1998b, 1999; Varney 2000; Probst and Konisky 2001: 75-76, 81-85).

State Roles

CERCLA provides for a substantial role for states in the Superfund program. Among the provisions involving states are requirements for a state to share costs for remedial actions (typically 10%), substantial and meaningful involvement in remedy selection, and the ability to carry out response actions. Specifically, CERCLA authorizes the federal government to enter into cooperative agreements with states and Indian tribes to carry out response activities consistent with the National Contingency Plan (CERCLA as amended by SARA, Section 104(d)(1)). EPA's regulations also authorize funding for building state programs to carry out those activities and to develop their own response programs. Since 1987, Superfund State Cooperative Agreements have totaled over $3 billion, including over $300 million for building and maintaining state programs. Consequently, states have performed a significant number of CERCLA site assessments, along with a much smaller number of RI/FSs, RDs, and RAs (in most circumstances, funding for removals from the CERCLA trust fund is reserved to EPA).

State agencies (as well as private firms) also respond to potential or actual releases of hazardous substances. The benefits of these responses are partially attributable to Superfund due to funding and technical assistance provided to states, the ability to use (or at least threaten to use) CERCLA's liability provisions, and the availability of information and technological innovations created by the Superfund program. Many, but not all, of these state and private responses are smaller and simpler than those handled by the federal government (Probst and Konisky 2001: 93-97). Further, state hazardous substance cleanup programs rely heavily on the federal Superfund program in a number of ways. The federal Superfund program has created the knowledge, technology, and skills needed to assess the risks of hazardous substance sites and clean them up safely. In addition, the existence of the Superfund law, with its very significant liability provisions, supports state programs, which can use the threat of federal enforcement actions to elicit cooperation from private firms. Moreover, under the Core State and Tribal Cooperative Agreements, the federal Superfund program has invested over $300 million to build and maintain state capabilities.

State response programs have grown in scope, capability, and sophistication, particularly over the last decade, and have provided an alternative to Superfund for many sites. Some states have developed the legal and technical capacities to deal with complex sites similar to those that typically end up on the NPL (General Accounting Office 1998a). States have been especially active in developing VCPs and programs to support brownfields redevelopment.

Liability and Private Firms

Many CERCLA responses involve the enforcement of CERCLA's liability provisions, in which EPA seeks to identify the potentially responsible parties (PRPs) – those individuals or organizations responsible for creating or contributing to uncontrolled releases of hazardous substances. CERCLA's two basic liability provisions permit EPA to either compel a PRP to abate an endangerment to public health, welfare, or the environment, or to recover the costs of response. The law also provides for citizen suits to enforce CERCLA's provisions (Section 310), and it provides authority for federal agencies, states, and tribes to bring actions for damages to natural resources (Section 107).

Liability can extend to site owners, facility operators, waste transporters, or anyone who generates hazardous substances that contaminate other sites. This liability is strict, joint, and several, with no requirement that a PRP's hazardous substance be the sole cause for a response action. Legal proof of negligence is not required, and conducting activities consistent with standard industry practices is not considered an adequate defense. The original draft of CERCLA contained no statute of limitations. This was altered in 1986 with SARA's inclusion of limits on recovery actions and natural resource damages penalties.

Federal Sites

A distinction is often made between federal and non-federal sites, where the former are typically former military facilities, national laboratories, and parts of the nuclear weapons complex. Examples include the former Fort Ord in Monterey, California, and the nuclear weapons facility at Hanford, Washington. Federal sites often have larger, more complex, and more challenging problems. The distinction of federal and non-federal is also important because of major differences in liability, response roles, and availability of the Trust Fund to pay for response. Moreover, federal facilities often are geographically large installations with numerous releases or potential releases that are somewhat separate and distinct from one another. Through FY 2003, a total of 175 federal sites were on the NPL, representing 11.3% of all sites.

WHAT SUPERFUND HAS ACCOMPLISHED SO FAR

In order to understand what the Superfund program has accomplished, it is useful to consider what the world would be like in the absence of the Superfund program (or a replacement). In this "without-Superfund" scenario, emergencies due to releases of hazardous substances might have continued to be handled as state and federal disasters, as occurred at Love Canal when the Federal Emergency Management Agency took charge. Moreover, without Superfund, the research, innovation, training, and enforcement supported by CERCLA and SARA would not have taken place. The benefits of these secondary impacts would be hard to estimate. For instance, capabilities developed with support from Superfund were crucial to recovering from the terrorist attacks in the fall of 2001, when, for example, the EPA Superfund program responded to anthrax contamination and monitored public and worker safety at the World Trade Center. In the without-Superfund scenario, the time and cost to recover from these attacks would likely have been much higher.

Similarly, without the enforcement activities and deterrence effects of Superfund, more uncontrolled releases of hazardous substances would likely have occurred, while the first responders who would have had to deal with the releases would have been less well prepared because they would not have benefited from Superfund-supported training.

Since the passage of CERCLA, many previously hidden instances of contamination have been discovered and new releases of hazardous substances have continued to occur. Most of these contaminations are located at either current or former industrial sites or waste disposal sites, but some are at military bases and facilities associated with nuclear weapons production.

The Superfund program deals both with places where releases have occurred due to deliberate actions (sites) as well as places where releases have been due to accidental actions (spills).

There is extremely little quantitative risk assessment data about removal actions, but qualitative information can be found in several sources. Several researchers have documented significant reductions in exposure and risk from exposure during both remedial and removal actions (von Lindern et al. 2003; Sheldrake and Stifelman 2003; Khoury and Diamond 2003). This is the only quantitative evidence that removal actions may significantly reduce risk.

Most analysts who have considered the question of whether removals mitigate significant real risks have come to the conclusion that they do (Koshland 1991; Hird 1994; Wildavsky 1995; General Accounting Office 1995; Office of Management and Budget 2003). For instance, Hird notes removals prominently in his section on "Superfund's Overlooked Accomplishments":

> Indeed, much of Superfund's success lies with the removal action program, which has removed more than 2,600 immediate threats to health and the environment since 1980 and has reduced substantial risks at many sites at relatively little cost. Thomas Grumbly, former Director of Clean Sites, Inc. and former U.S. Department of Energy Assistant Secretary for Environmental Management, stated that EPA's removal program, "has probably eliminated most of the *immediate* health risks posed by abandoned hazardous waste sites" (Hird 1994: 29-30).

There are some qualitative comparisons between removal and remedial actions: "these emergency activities may operate in tandem with, *or as replacements for*, the remedial process" (Hird 1994: 19; emphasis added). An OTA report noted that removal actions can resemble a remedial cleanup (Office of Technology Assessment 1989: 7, box). These comparisons are not quantitative, but they provide support for the idea that some removal actions result in significant health benefits.

Without the intervention of Superfund, the magnitude of such effects likely would have worsened over time, as more and more containers and facilities holding hazardous substances failed, as leaked substances spread through ground water, and as more people came to live near or even on such sites.

It is important to recognize that a crucial part of the hazardous substances problem in 1980 was that very little was known about the nature or extent of the problem. While there were indications that hazardous substances had contaminated many places throughout the country, and it was known that some of these substances had physiological effects, there was a great deal of uncertainty as to the number of such problems and the nature and magnitude of the associated risks to human health and the environment. There was also very little knowledge about how best to remediate contaminated sites. This lack of knowledge is unsurprising, given the laws and incentives up to 1980; there was no reason for private industry to invest in these scientific and engineering questions, and before the existence of a public policy problem was identified, little reason for government to sponsor such research. However, this lack of knowledge created uncertainty and concern among the public about the potential impacts of hazardous substance releases on the health and well-being of their families. The Superfund program has greatly reduced the uncertainty associated with the problem of uncontrolled releases of hazardous substances and provided much better tools to manage the problem.

FUTURE CHALLENGES FOR SUPERFUND

While much has been learned about better management of hazardous substances since CERCLA was passed in 1980, there are still many challenges for the Superfund program. In particular, many of the largest, most challenging NPL sites remain to be dealt with, including many of the massive federal sites. The number of contaminated sites across the country and the difficulties associated with dealing with them are more substantial than was originally suspected. The process for identifying, studying, and cleaning up contaminated sites (e.g., the NPL pipeline) has often proved to take a long time, cost a great deal of money, and be very contentious. This has provoked concern and resentment among PRPs, who often see themselves as paying to clean up someone else's mess. There is also a widespread perception that NPL remediation reduces risks only very slightly, on average, for very large costs, and that even these risks are largely hypothetical (Viscusi and Hamilton 1999). However, this view may ignore almost all of the risk mitigation that the Superfund program has accomplished, due to a narrow focus on the NPL (and not on removals, which appear to be the cause of most of the risk mitigation) and a lack of data about key health impacts.[7] Nonetheless, this view is prevalent.

Further, one of the key challenges for the Superfund program, and indeed for all efforts to deal with releases of hazardous substances, becomes clear if sites are considered as contaminated properties and the potential for the use of

---

[7] Hamilton and Viscusi focus on cancer and ignore other health effects due to data unavailability, although the evidence suggests the majority of the health risk is due to these other health effects, especially birth defects (Johnson 1999; Dolk and Vrijheid 2003; Khoury and Diamond 2003).

these properties in the future is considered. This is an increasingly important concern as more NPL sites are remediated and as the interest in restoring abandoned "brownfield" sites increases. It is not clear that current approaches to managing releases of hazardous substances are up to this task, suggesting a role for new approaches such as the application of adaptive management described in the remainder of this volume.

REFERENCES

Agency for Toxic Substances and Disease Registry. 2003a. *ATSDR CEP Site Count Report.* Atlanta: ATSDR Division of Toxicology.

———. 2003b. *CERCLA Priority List of Hazardous Substances.* Atlanta: ATSDR.

Anderson, B., C. Thompson, and W. A. Suk. 2002. "The Superfund Basic Research Program - Making a Difference: Past, Present, and Future." *International Journal of Hygiene and Environmental Health* 205 (1-2): 137-141.

Andrews, R. N. L. 1999. *Managing the Environment, Managing Ourselves.* New Haven, CT: Yale University Press.

Association of State and Territorial Solid Waste Management Officials. 1998. *State Cleanup Accomplishments for the Period 1993-1997.* Washington, DC: ASTSWMO.

Baumol, W. J., and W. E. Oates. 1988. *The Theory of Environmental Policy.* 2nd ed. New York: Cambridge University Press.

Bove, F., Y. Shim, and P. Zeitz. 2002. "Drinking Water Contaminants and Adverse Pregnancy Outcomes: A Review." *Environmental Health Perspectives* 110: 61-74.

Brown, R. S. 2001. *States Put Their Money Where Their Environment Is.* Washington, DC: Environmental Council of the States (ECOS).

Congressional Budget Office. 1994. *The Total Costs of Cleaning Up Nonfederal Superfund Sites.* Washington, DC. : Congressional Budget Office.

Dolk, H., and M. Vrijheid. 2003. "The impact of environmental pollution on congenital anomalies." *British Medical Bulletin* 68: 25-45.

Environmental Law Institute. 1998. *An Analysis of State Superfund Programs: 50 State Study, 1998 Update.* Washington, DC.

———. 2002. *An Analysis of State Superfund Programs: 50 State Study, 2001 Update.* Washington, DC.

EPA Science Advisory Board. 1990. *Review of the Superfund Innovative Technology Evaluation (SITE) Program.* Washington, DC: Environmental Protection Agency.

Fung, A., and D. O'Rourke. 2000. "Reinventing Environmental Regulation From the Grassroots Up: Explaining and Expanding the Success of the Toxics Release Inventory." *Environmental Management* 25 (2): 115-127.

General Accounting Office. 1995. *Superfund: Information on Current Health Risks.* Washington, DC: General Accounting Office.

———. 1998a. *Hazardous Waste Sites: State Cleanup Practices.* Washington DC: General Accounting Office.

———. 1998b. *Hazardous Waste: Unaddressed Risks at Many Potential Superfund Sites.* Washington, DC: General Accounting Office.

———. 1999. *Superfund: Progress, Problems, and Future Outlook.* Washington, DC: General Accounting Office.

Hamilton, J. T. 1995. "Pollution as News - Media and Stock-Market Reactions to the Toxics Release Inventory Data." *Journal of Environmental Economics and Management* 28 (1): 98-113.

Hamilton, J. T., and W. K. Viscusi. 1999. "Calculating Risks? The Spatial and Political Dimensions of Hazardous Waste Policy." In *Regulation of Economic Activity*, N. L. Rose and R. Schmalensee, eds. Cambridge: MIT Press.

Harrison, R. M. 2003. "Hazardous Waste Landfill Sites and Congenital Anomalies: Where Do We Go From Here?" *Occupational and Environmental Medicine* 60 (2): 79-80.

Hays, S. P. 1987. "Beauty, Heath, and Permanence: Environmental Politics in the United States, 1955-85." In *Studies in Environment and History*, D. Worster and A. Crosby, eds. New York: Cambridge University Press.

Hird, J. A. 1994. *Superfund: The Political Economy of Environmental Risk*. Baltimore: Johns Hopkins University Press.

Johnson, B. L. 1999. *Impact of Hazardous Waste on Human Health: Hazard, Health Effects, Equity, and Communications Issues*. Boca Raton: Lewis Publishers.

Jung, C., K. Krutilla, and R. Boyd. 1996. "Incentives for Advanced Pollution Control Abatement Technology at the Industry Level: An Evaluation of Policy Alternatives." *Journal of Environmental Economics and Management* 30 (1): 95-111.

Khoury, G. A., and G. L. Diamond. 2003. "Risks to Children from Exposure to Lead in Air During Remedial or Removal Activities at Superfund Sites: A Case Study of the RSR Lead Smelter Superfund Site." *Journal of Exposure Analysis and Environmental Epidemiology* 13 (1): 51-65.

King, G. 2002. Letter to Michael B. Cook, Director of the Office of Emergency and Remedial Response. Washington, DC: Association of State and Territorial Solid Waste Management Officials.

Konar, S., and M. A. Cohen. 1997. "Information as Regulation: The Effect of Community Right to Know Laws on Toxic Emissions." *Journal of Environmental Economics and Management* 32 (1): 109-124.

Koshland, D. E. 1991. "Toxic-Chemicals and Toxic Laws." *Science* 253 (5023): 949-949.

Landy, M. K., M. J. Roberts, and S. R. Thomas. 1994. *The Environmental Protection Agency: Asking the Wrong Questions from Nixon to Clinton*. New York: Oxford University Press.

National Conference of State Legislatures. 2004. "Superfund Reform Policy." Washington, DC: NCSL.

National Governors Association. 2003. "Policy Position: Superfund Policy." Washington, DC.

Office of Emergency and Remedial Response. 1989. *Risk Assessment Guidance for Superfund*. Volume I, Human Health Evaluation Manual (Part A). Washington, DC: Environmental Protection Agency.

———. 1990. *National Contingency Plan*. Washington, DC: Environmental Protection Agency.

Office of Management and Budget. 2003. *Performance and Management Assessments*. Washington, DC.

Office of Solid Waste and Emergency Response. 1990. *Superfund Removal Procedures Action Memorandum Guidance*. Washington, DC: Environmental Protection Agency.

———. 1998. *RCRA, Superfund & EPCRA Hotline Training Module: Introduction to Superfund Accelerated Cleanup Model*. Washington, DC: Environmental Protection Agency.

Office of Technology Assessment. 1989. *Coming Clean: Superfund Problems Can Be Solved*. Washington, DC: U.S. Congress.

Orr, L. 1976. "Incentive for Innovation as the Basis for Effluent Charge Strategy." *American Economic Review* 66 (2): 441-447.

President of the United States. 1994. Executive Order 12898: Federal Actions to Address Environmental Justice in Minority and Low-Income Populations.

———. 1999. Executive Order 13132: Federalism.

———. 2000. Executive Order 13175: Consultation and Coordination with Indian Tribal Governments.

Probst, K. N., D. Fullerton, R. E. Litan, and P. R. Portney. 1995. *Footing the Bill for Superfund Cleanup: Who Pays and How?* Washington, DC: Brookings Institution and RFF Press.

Probst, K. N., and D. M. Konisky. 2001. *Superfund's Future: What Will It Cost?* Washington, DC: RFF Press.

RCRA/CERCLA Division. 1993. *Remedial Investigation/Feasibility Study (RI/FS): Process, Elements and Techniques.* Edited by Office of Environmental Guidance. Washington, DC: U.S. Department of Energy.

Sheldrake, S., and M. Stifelman. 2003. "A Case Study of Lead Contamination Cleanup Effectiveness at Bunker Hill." *Science of the Total Environment* 303 (1-2): 105-123.

Taylor, M. R., E. S. Rubin, and D. A. Hounshell. 2003. "Effect of Government Actions on Technological Innovation for SO2 Control." *Environmental Science & Technology* 37 (20): 4527-4534.

Traceski, T. T. 1994. *Environmental Guidance on CERCLA Removal Actions.* Washington, DC: Office of Environmental Guidance.

————. 1995. *Baseline Risk Assessment: Toxicity and Exposure Assessment and Risk Characterization.* Washington, DC: Office of Environmental Policy and Assistance.

Statement of the Commissioner of the New Hampshire Department of Environmental Services on Superfund on Behalf of the Environmental Council of the States. Washington, DC.

von Lindern, I., S. Spalinger, V. Petroysan, and M. von Braun. 2003. "Assessing Remedial Effectiveness Through the Blood Lead: Soil/Dust Lead Relationship at the Bunker Hill Superfund Site in the Silver Valley of Idaho." *Science of the Total Environment* 303 (1-2): 139-170.

Wildavsky, A. 1995. *But Is It True? A Citizen's Guide to Environmental Health and Safety Issues.* Cambridge: Harvard University Press.

2

ADAPTIVE MANAGEMENT IN SUPERFUND:
Thinking like a Contaminated Site

Jonathan Z. Cannon
University of Virginia School of Law

INTRODUCTION

Over the last three decades adaptive management has emerged as one of the most promising innovations in natural resource management and environmental regulation. Yet the possible benefits of this approach for Superfund, which is among the nation's most expensive and controversial environmental programs, have not been comprehensively explored. A 2003 study by the National Research Council (NRC) represented the first serious effort to apply adaptive management principles to cleanup of contaminated sites, with specific attention to contaminated Navy facilities under Superfund, the Resource Conservation and Recovery Act, and state regulatory statutes (NRC 2003; NRC et al. 2003). This chapter examines adaptive management for Superfund as a whole, including the privately owned sites that predominate within the Superfund universe. It elaborates the principles of adaptive management, explains how these principles might work within the legal and policy framework of Superfund, and explores their implications for managing individual Superfund sites as well as for administering the entire inventory of these sites. In the process, it sheds further light on the potential usefulness of adaptive management, which was developed for management of complex natural ecosystems, for a program dealing with local site contamination in largely urban settings.

The chapter concludes that, in the complex and uncertain world within which it must operate, Superfund does have something to learn from adaptive management. Superfund would work better, adaptive management principles suggest, with five changes in the framing and management emphasis of the Superfund program:

1. EPA should adopt a broad and flexible view of the public interest affected by Superfund sites. This expanded notion of the public good would encompass not only the values made explicit in

the Superfund statute, such as environmental protectiveness, but also other values that emerge from consultation with those most affected by a site's disposition. It would give future use of sites a central importance in the Agency's decisions.

2. EPA should promote and monitor institutional innovations, including collaborative stakeholder processes, to clarify and order values in deliberations on alternate futures for a site.

3. In the lengthy process of site study, remediation, and post-remedial review, EPA should improve monitoring and feedback mechanisms focused on crucial unknowns or uncertainties at a site and revisit and adjust prior decisions as warranted in light of new information. In particular, the Agency should improve its information gathering and review of anticipated future uses of a site in tandem with its planning, implementation, and review of cleanup actions.

4. Acknowledging the ability of players in both the public and private sectors and at multiple levels of government to affect outcomes at a site, EPA should foster the integration of decisions across sectors and jurisdictional scales.

5. EPA should employ conscious policy learning in its management of the entire portfolio of sites. It should consider framing program policies on controversial issues or questions involving scientific or technical uncertainty as experiments and commit to systematic recording and analysis of program experience as a basis for review and change.

More generally, the chapter recommends that the Agency embrace adaptive management principles in administering Superfund. Superfund as currently implemented, including recent Agency initiatives in several of the areas mentioned above (Cook 2005a; EPA 2004b; Davies 2001), provides some support for an adaptive approach. But "adaptive management has not yet been incorporated into the [cleanup] process as a whole" (NRC 2003: 4), nor has Superfund adopted it as a management guide. Systematic application of adaptive management principles will be necessary to realize the full potential of this approach.

A Model of Superfund

The chapter bases its analysis on a model of Superfund as a program for the management of contaminated sites by multiple parties, over extended time periods, and across a range of values or policy objectives. This model differs from what has been the prevailing concept of Superfund as time-limited intervention by federal officials focused predominantly on public health concerns. More specifically, the model challenges three major aspects of Superfund, as traditionally understood. The first of these is that contaminated sites can be dealt with effectively and decisively over a relatively limited time: the problem is defined, a remedy is ordered and implemented, and the problem is solved (i.e., the site is "cleaned up") (NRC 2003: 3). The statute (42 U.S.C. § 9621(b)(1) (2000)) states a preference for treatment remedies that "permanently" reduce the volume and toxicity of contaminants and the early emphasis of both EPA and its congressional overseers was on rapid and complete cleanups. But, as it turned out, Superfund cleanups took much longer than initially anticipated, and, even more significantly for our purposes, most Superfund sites have contaminants remaining after the remedy is completed and will require long-term monitoring and review. At least for these sites with lingering contaminants, a more accurate program model is one in which site interventions by EPA are understood to occur in multiple phases over an extended period of time and under conditions of uncertainty and change. This is a model for which adaptive management, with its focus on experimental action and continuous learning through program monitoring and evaluation, is particularly suited.

Second, the Superfund model advanced here addresses the program's historical focus on protecting public health. This focus has been effective in forcing action to reduce health risks posed by site contamination, but has excluded or marginalized consideration of other value-significant dimensions of these sites, such as ecosystem function, economic development, and compatibility with community norms and aspirations. Superfund sites are resources (e.g., land and associated groundwater) whose restoration and future use may offer substantial economic benefits or advance important community values, including but not limited to the reduction of health risks. Although the Superfund statute makes "protectiveness" the central consideration in cleanup decisions, there is considerable flexibility in what "protectiveness" requires. The statute also makes room for consideration of other objectives or values, and indeed these other values have made their way into the Superfund decision process, although often not directly or explicitly, as felt necessities of these sites. This chapter argues for an expansive deliberative scope for Superfund, one in which a range of potentially competing values are brought to bear in deliberations over alternative site futures. Adaptive management encourages institutional innovations, including collaborative stakeholder processes, by which such values can be identified, ordered, and applied. This management model presumes that relevant preferences are, at least to some degree, endogenous to the deliberative process, and emerge as stakeholders absorb scientific and technical information in considering alternative scenarios for cleanup and reuse.

The third program feature addressed by the model is the dominant federal role in decision-making at Superfund sites. The Superfund statute places the authority for remedial decisions at National Priority List sites in the hands of federal officials, and that authority, unlike site-specific decisional authority under many other federal environmental statutes, is not delegable to the states or local jurisdictions. States often have concurrent authority under their own laws for cleanup of contaminated sites, and in some cases EPA defers to state cleanup decisions under these laws. But more importantly for our purposes, the selection of a remedy is only one of the decisions that determine the future of a Superfund site. Subject to constraints imposed by a remedy selected by EPA, property owners and local authorities typically determine site use, and future use of a site can significantly impact remedy selection, implementation, and effectiveness over time. Thus, decision-making at Superfund sites is best understood as a series of interactions among actors operating at multiple levels of government and across the public and private sectors. Adaptive management, with its emphasis on systems hierarchies, provides a framework for the integration of decision-making across scales and seems particularly suited to the multi-scalar complexity of the Superfund program.

A Snapshot of the Superfund Program

Superfund sites represent a relatively small portion of the universe of contaminated sites in this country, but they include many if not most of the largest, environmentally most problematic, and politically most contentious sites, and thus have a policy importance disproportionate to their number (EPA 2005c, 2005l). Although much progress has been made on the current inventory of Superfund sites, much work remains to be done. As of October 2005, EPA had placed 1547 sites on Superfund's National Priorities List (NPL) (EPA 2005e). Of these, 308 had been deleted from the NPL, leaving 1239 on the current list. Of the total 1547 sites, construction of the remedy was complete at 966 (or about 60 percent). The term "construction complete" means that all physical construction for remedies at these sites is complete, but does not mean that long-term cleanup goals have been met (Federal Register 1990). Only 248 of the 1547 NPL sites, less than one-third of the "construction complete" sites, were in productive use (EPA 2005k). Estimates of the total universe of contaminated sites in this country have ranged between 70,000 and 500,000 (Probst et al. 2001: 85). A study by Resources for the Future has projected that, each year over the next decade, between twenty and fifty of these sites will be added to the NPL and many more will become subject to action under state cleanup programs (Probst et al. 2001: 105). Although this chapter focuses on Superfund, the principles it develops are also applicable to state programs, including cleanup and reuse of brownfield sites.

ADAPTIVE MANAGEMENT: LEARNING TO THINK LIKE A MOUNTAIN

In his canonical essay, "Thinking Like A Mountain," Aldo Leopold describes a process by which he came to understand the critical role of predators in ecosystems (1949: 129). Rather than producing a "hunters' paradise," he discovered that eliminating wolves led to an overpopulation of deer, the depletion of foliage, and the degradation of the mountain ecosystem on which the deer depend. Leopold observed a similar effect from ranchers' clearing range land of wolves, which resulted in overgrazing. According to Leopold, the ranchers responsible for such clearing have not "learned to think like a mountain. Hence we have dustbowls, and rivers washing to the sea" (132). "We all strive for safety, prosperity, comfort, long life, and dullness," Leopold observed, "but too much safety seems to yield only danger in the long run" (133). At first glance, nothing could be more different from Leopold's mountain than a contaminated site. One represents a relatively undisturbed ecosystem, the other a highly disturbed environment, a place that is far from its "natural state"; over 80 percent of listed Superfund sites are in Metropolitan Statistical Areas (E^2 Inc. 2005: 3-35). Yet, although highly modified by human activity, Superfund sites do contain natural systems or are embedded within natural systems that provide important resources and services, such as supplying groundwater and surface water and providing habitat for humans and other life forms, from soil microbes to rare megafauna. These sites are also located within human socio-economic systems, which are closely linked to natural systems. Urban ecologists argue that biophysical and socio-economic systems in city landscapes should be considered together as parts of a single "urban ecosystem" and that the study of these systems should integrate consideration of biophysical and socio-economic drivers, including human values and institutions (Lord et al. 2003: 320, 324-26).

Like the management of Leopold's wolf-deer-cow ecosystem, the management of Superfund sites and the human-natural systems within which they are situated has future implications for the welfare of humans and other living things. Thinking like a contaminated site means accepting stewardship obligations and employing new learning, just as Leopold does in coming to think like a mountain. Thinking like a contaminated site is the province of adaptive management (Norton and Steinemann 2001: 487).

Adaptive management has its origins in ecosystem management, where scientists and policymakers have encountered high degrees of uncertainty (and surprise) in the dynamics of natural systems and in the responses of those systems to human intervention (Holling 1978; Walters 1986; Lee 1993). It developed as a strategy of institutional "learning while doing." In adaptive management, policy decisions have a provisional, experimental quality; decision-makers maintain flexibility and adjust as they go, based on monitoring the effects of their past decisions and on new information from other sources. A National Research Council study describes adaptive management as "involv[ing] a decision-making process based on trial, monitoring and feedback. . . and recogniz[ing] the imperfect knowledge of interdependencies existing within and among natural and social systems, which requires plans to be modified as technical knowledge improves" (NRC 1992: 357).

Proponents distinguish adaptive management from "old-fashioned 'trial and error,' a crude and familiar process in which the manager simply tries an approach thought most likely to succeed, and if it fails, moves on to the next most likely successful alternative" (Karkkainen 2003: 949, citing Walters 1986: 64). Kai Lee advocates an "active" form of adaptive management, in which policies are explicitly selected and designed as experiments to ensure that "the most important uncertainties are tested rigorously and early" (Lee 1999: 3). Less rigorous forms of adaptive management, which avoid the explicit framing of policies as experiments, include providing mechanisms for regular collection and feedback of information, specifying points for policy review and re-evaluation, and maintaining openness and flexibility (Walters 1986: 248-252; Torrell 2000: 353-354; Meffe et al. 2002: 103, 106; Moir and Block 2001: 141-142). Even these more modest versions of adaptive management provide significantly greater capacity for learning than the traditional reactive management approaches.

These strategies for learning can extend not only to scientific but also to socio-economic and political uncertainties and can support the evolution of process as well as policy. Adaptive management envisions a dialectic between technical and scientific information and values affecting policy choices. Its earliest proponent, C. S. Holling, and co-author Stephen Light have stated that the focus of adaptive management should be on the "coupled dynamics of nature, society and resource institutions" (Gunderson, Holling, and Light 1995: 508). Kai Lee characterizes adaptive management as "[l]inking science and human purpose" (1999: 9). But often neither the science nor the human purpose is clear: there is "disagreement over both means and ends" (105). Lee proposes a collaborative process in which stakeholders frame the questions to be answered through policy experimentation, and also negotiate among themselves on the relative desirability of outcomes, selecting among alternative futures or development paths (Lee 1993: 104-114; cf. Norton & Steinemann 2001: 478; McDaniels & Gregory 2004: 1921-1922). To advance the goals of adaptive management, these processes must be established at the outset of policy deliberations; must be ongoing or iterative; and must have learning as a central objective (McDaniels & Gregory 2004: 1921).

Adaptive management proponents emphasize the hierarchical qualities of the human-natural systems they address. Systems are nested, operating at different spatial and temporal scales, and the linkages between these systems are a source of additional complexity and uncertainty (Norton & Steinemann 2001: 479). This multi-scalar perspective applies not only to physical and biological processes with "distinct frequencies in space and time" (Holling 1992: 447-48) but also to institutional systems operating at different levels and sensitive to different interests and values (Norton & Steinemann 2001: 478-79). Bryan Norton and Anne Steinemann incorporate axioms of hierarchy theory into the understanding of adaptive management, as a "means to organize the spatial and temporal relationships that are so important in multi-scalar management." In particular, the first axiom of hierarchy theory ("all observation and measurement must be oriented from some point within the system") "operationalizes both a scientific and political focus *from a specific locale, which represents a point within a complex, dynamic, and multi-scalar system*" (emphasis in original). Accordingly, Norton and Steinemann argue for a place-based or community-based approach, in which "inputs" from local groups "serve as a starting point in the search for management goals" (2001: 480-82). A key challenge of adaptive management in the Superfund program is to distinguish appropriate roles for federal, state, and local players and to design processes to integrate their distinct perspectives into a coherent, adaptive policy framework.

Superfund sites present the sorts of uncertainties and opportunities for learning over extended periods for which adaptive management is particularly suited. Decisions require information about (1) the nature, quantity, and location of contaminants on site; (2) site characteristics, including ecosystem processes such as groundwater flow and microbial activity; (3) costs and effectiveness of remedies; (4) political and economic conditions affecting cleanup and reuse; and (5) values affecting the merits of alternative site futures. Studies and other information-gathering exercises are undertaken to obtain this information, but significant uncertainties often remain after the studies are complete, and new information is generated throughout the cleanup process. Some of this information may come by way of response to decisions made and implemented or sought to be implemented at the site. For example, an attempt to carry out a groundwater pump and treat remedy may reveal new information about cost or effectiveness that would cause decision-makers to rethink the remedy. Adaptive management sees these sites and the human-natural systems in which they are situated as dynamic, unfolding at multiple scales of space and time.

Superfund also obviously lends itself to management that is oriented to a particular place or resource—the site—and that emphasizes deliberation among stakeholders identified particularly with that site. Compared to other environmental programs in which detailed regulatory standards drive decisions toward uniform results across diverse environmental and socioeconomic settings, Superfund's decision criteria for cleanups, as articulated in the Comprehensive Environmental Response, Compensation, and Liability Act (CERCLA) and the National Contingency Plan (NCP), are relatively open and flexible. Thus there is room to tailor decisions according to both the physical circumstances of the site (as well as the larger systems to which the site is connected) and the preferences or values of the community around it (as well as those of stakeholders at the state or national level).

Adaptive management in Superfund can be applied on a site-by-site basis and also a site portfolio basis, as the Agency adjusts its management of its entire inventory of sites or distinct portions thereof in light of its program experience and other sources of new information. I explore the individual site applications first.

DELIBERATIVE SCOPE/DELIBERATIVE PROCESS: CLARIFYING HUMAN PURPOSE

The early implementation of Superfund heavily emphasized reducing the human health risks posed by contaminated sites to acceptable levels. Public debate focused on how quickly and effectively EPA was achieving that end, and for the most part it still does (Dunne 2004). As the program has matured, however, it is evident that other concerns and values have important bearing on the disposition of these sites, including values expressed through the market (e.g., market efficiency), through local government processes (e.g., community welfare), and through other non-market institutions such as local environmental groups (e.g., ecological sustainability) or neighborhood associations (e.g., neighborhood identity or amenities) ("Green Light" 2002; Front Royal 2005). Decisions by local stakeholders acting on these values can affect the long-term stability and effectiveness of federal remedy decisions, and federal remedy decisions can affect the ability of these stakeholders to realize these values in the site's ultimate disposition.

This chapter argues for the explicit recognition, in federal decision-making, of the multiplicity of values at play in determining actual outcomes at these sites, and in the section below analyzes the extent to which Superfund, as currently written, can accommodate such recognition. This deliberative breadth is a necessary implication of the linkage between "science and human purpose" that is at the heart of adaptive management. Failure to take account of a full range of values weakens the linkage, undermining the legitimacy of both the process and its outcomes among those most affected (Jiggins & Röling 2002: 97). Because these values are not well defined at the outset of the cleanup process, this chapter further urges EPA to increase its support for collaborative stakeholder processes that integrate the clarification of relevant values with the evaluation of alternative site futures (Lee 1993: 105; Lindblom 1959: 81-83).

Deliberative Scope: Superfund Decision Structure

All Superfund remedies must meet applicable or relevant and appropriate requirements ("ARARs") under other federal and state environmental laws and must also achieve EPA's more general requirement of "overall protection of human health and the environment." EPA has designated these two requirements as "threshold criteria," and this chapter will sometimes refer to them together as "protectiveness." EPA's regulations identify seven other decision criteria for Superfund remedy selection. Five of these are "balancing criteria": long-term effectiveness and permanence; reduction of toxicity, mobility, or volume through treatment; short-term effectiveness; implementability; and cost. The two remaining factors—acceptability of the remedy to the state and to the local community—are "modifying criteria." None of the balancing or modifying criteria may override the core requirement of protectiveness.

The general protectiveness criterion addresses the environmental benefits of cleanup. Environmental benefits include gains that relate directly to human health and those that do not, such as restoration of ecosystem services and protection of biodiversity. A 1995 study of Superfund remedial decisions found an "almost exclusive" reliance on human health considerations, to the exclusion of ecological concerns (Walker et al. 1995: 29; cf. Suter et al. 2000), but the Agency states that at present a number of its major cleanup actions are driven by ecological protectiveness (Cook 2005b). With the exception perhaps of cost and implementability as "balancing criteria," the statute and the regulations make no express provision for the consideration of the non-environmental values that might be implicated in competing remedial alternatives. There is considerable play, however, in what protectiveness requires. For example, for carcinogenic contaminants, EPA's regulations require cleanup to an individual excess cancer risk of no more than 10^{-4} (one in 10,000) to 10^{-6} (one in 1,000,000). Thus, the permissible residual cancer risk after cleanup varies by two orders of magnitude. Moreover, the assessment of risk will vary widely depending on the assumptions made about such factors as actual and potential exposure to hazardous substances at the site, which—as we shall see—will vary with projections of future use and other conditions affecting the site. There is even broader discretion in determining the level of protectiveness required for ecological risks (Luftig 1999). EPA also has discretion in the application of ARARs, including the authority to waive ARARs under certain circumstances (see United States v. Akzo Coatings 1991: 1409, 1446-1450).

It is in this realm of discretion that a broader consideration of the public good may take place, including consideration of the non-environmental values associated with alternative remedial or reuse scenarios. These values may enter the deliberations in at least two contexts. The first is the Agency's consideration of "reasonably anticipated future land use," under guidance issued for the preparation of the Remedial Investigation/Feasibility Study (RI/FS), discussed below. The second is the statutory requirement that EPA consider the views of the community on its proposed remedy. Both are discussed below.

Reasonably Anticipated Future Land Use

As part of the remedial investigation, EPA conducts a baseline risk assessment, which includes the determination of a "reasonable maximum exposure [to contaminants on site] expected to occur under both current and future land use conditions" (EPA 1989). In a May 1995 guidance, the Agency stated that the "[f]uture use of the land will affect the types of exposures and the frequency of exposures that may occur to any residual contamination remaining on site, which in turn affects the nature of the remedy chosen" (Land Use Directive 1995; Hersh et al. 1997). The guidance requires that cleanup objectives reflect "reasonably anticipated future land use" and states further that "[l]and uses that will be available following completion of remedial action are determined as part of the remedy selection process" (Land Use Directive 1995: 2). Thus, the reasonably anticipated future land use becomes embedded in the remedial decision and may exclude certain future reuse options while facilitating others. EPA has recently acknowledged this effect of the future land use determination, stressing the importance of "[i]ntegrating realistic assumptions of future land use into Superfund response [as] an important step toward facilitating the reuse of sites following cleanup" (Reuse Assessment 2001).

The 1995 guidance concerning land use came in response to criticisms that the Agency's risk assessments and remedial decisions had reflexively assumed the future use would be residential. This assumption, critics argued, raised projected levels of exposure to contaminants left on site, leading to more aggressive cleanup objectives and more expensive remedies (Land Use Directive 1995: 3). Although the directive does not explain it this way, by considering "reasonably anticipated future land use," EPA can avoid remedies whose incremental costs would not be justified by the incremental benefits, such as an aggressive remedy that made a site "safe" for residential use when only industrial use was likely.

Consideration of future land uses allows, in some rough sense, a comparison of the benefits as well as the costs of various remedial/reuse options. Each potential land use will be associated with potential benefits (e.g., profits to the site owner, increases in neighboring property values, and contributions to community values not fully captured in property values) that are in addition to the human health and environmental benefits flowing from the cleanup. Each land use scenario will also be associated with redevelopment costs, in addition to the expense of the cleanup that would be required to support it.

Under the current guidance, EPA does not represent itself as selecting a "reasonably anticipated future land use" based on a determination of what would be in the overall public interest (a preference or value judgment); instead it sees itself as developing realistic assumptions about what land use will occur (a factual determination). However, the relative likelihood of a selection among various reuse options reflects a determination by someone or some institution (e.g., the owner, the real estate market, neighborhood groups, zoning officials, or some combination thereof) that, in the particular circumstances of a site, some uses are preferable to others. The process of developing these assumptions about future land use contains an implicit value judgment.

Moreover, EPA does not automatically accept a "reasonably anticipated land use" proffered by the community, but must balance this preference for future land use with other technical and legal considerations provided in the Superfund law and its implementing regulations (Land Use Directive 1995: 7; Reuse Assessment Guide 2001: 2). Specifically, EPA balances the requirements to treat principal threats, to use engineering controls such as containment for low-level threats, to use institutional controls to supplement engineering controls, and to consider the use of innovative technologies (Land Use Directive 1995: 7). In addition, EPA must comply with other laws when they are "applicable or relevant and appropriate" (EPA 2001b: 14). Thus, EPA adds its own review to the community preferences reflected in the "reasonably anticipated future land use."

In sum, consideration of land use can provide indications of how market efficiency and other values to the community may be affected by various site management options. When combined with other information within EPA's consideration, information about future land uses will aid selection of a remedial/reuse option that is in the overall public interest. In an era of limited federal cleanup dollars and possible reduced participation in cleanups by responsible parties, it may also help attract funding for cleanup from prospective developers (Dunne 2004: 5, 8).

Ideally, the agency's remedial decision will facilitate a cleanup/reuse package that enhances the site's value within the statutory constraints of protectiveness. It will be the agency's responsibility to manage sites in ways that encourage, or at least do not foreclose, beneficial long-term strategies. This is not easy, for several reasons. First, determining "reasonably anticipated future land use" may be very difficult during the remedial decision-making process. In particular, "[a]t nearly 80% of sites on the NPL, there are adjacent residential areas." Thus, "[p]redicting the 'future land use' of these sites could be difficult" (Hersh et al. 1997: 70, citing EPA 1995b). Second, even where "reasonably anticipated future land use" can be determined, that determination may only be possible in broad categories, such as industrial, commercial, recreational, or ecological. Indeed, EPA anticipates that the reuse assessment will be documented in these broad terms and that "[m]ore specific end uses (e.g., office complex, shopping center, or soccer facility) can be considered during the response process when detailed planning information is readily available" (Reuse Assessment Guide 2001: 1). A study published by Resources for the Future found that "anticipated use of a site often evolves in tandem with the site remedy" (Hersh et al. 1997: 6). It may take years for reuse plans to take on specific form, if they emerge at all. Third, the "reasonably anticipated land use" may change. Site ownership, market conditions, and political alignments within the local jurisdiction with land use control authority over the site can all change unpredictably, with direct implications for future land use.

Despite these difficulties, there are several steps that EPA could take to improve the accuracy and usefulness of reuse assessments over the long term, particularly where those assessments are likely to have determinative effect on remedy selection, design, or implementation. The first step would be to reduce uncertainty by investing in more vigorous examination of future land use options in the Remedial Investigation/Feasibility Study (RI/FS) phase. Current land use guidance emphasizes that the reuse assessment should "rely on readily available information" (Reuse Assessment Guide 2001: 1). Given the role of future land use envisioned here (i.e., capturing important public values that might otherwise be missing from site deliberations and reflecting particularly the concerns and preferences of the community most immediately affected by EPA's decisions), going beyond "readily available information" will likely be warranted. Second, the Agency should take affirmative steps to enhance the likelihood that the "reasonably anticipated future land use" has institutional support within the local jurisdiction. In determining reasonable future land use, EPA is to consult with local land use planning authorities, local officials, and the public (Land Use Directive 1995: 4) or solicit "community input" (Reuse Assessment Guide 2001: 7). But more than "consultation" may be appropriate. The individuals with whom EPA consults might be expected to have a diversity of views on what should be done with a site. Reconciling those views may warrant EPA sponsorship of focused deliberation on the future of the site, rather than the more passive inquiry contemplated by the guidance, and, if resolution is achieved, the memorialization of results in contractual commitments and/or planning and zoning measures may be advisable (Davies 2001: 3). Finally, particularly where significant uncertainty about future land use remains, EPA should retain flexibility for adjustments in the remedy or its implementation in response to the emergence or refinement of promising reuse proposals.

A recent pilot project funded by EPA and conducted by the Hagerstown Land Use Committee, E^2 Inc., and the University of Virginia's Institute for Environmental Negotiation offers an example of an intensive community-based process to elicit future land use preferences serving the decision needs addressed above (Hagerstown Land Use 2003). At the Central Chemical site in Hagerstown, Maryland, a Land Use Committee, sponsored by Hagerstown's Planning Department, was convened, meeting a half dozen times among themselves and three times with the general public. The Committee's eighteen members included residents and property owners from around the site and from the city-at-large, local business interests, and government officials, and the site owner and other potentially responsible parties (Hagerstown Land Use 2003: 34). Expertise was also provided by "resource members," including representatives of the Planning Department, Hagerstown's Fire Department, and Maryland's Department of the Environment. Among the "guiding principles" or values expressly incorporated into the committee's deliberations were to "[p]rotect the long-term health and safety of community residents"; "[e]nsure that site reuses are compatible with surrounding neighborhoods"; "provide community-wide benefits," including the creation of tax benefits and new jobs; "integrate the natural environment into the site's reuse"; and "[u]nderstand the site within its local surroundings and as part of the larger community" (2003: 6).

The committee reached consensus, recommending that the site be reused for either mixed light industrial development (with a natural buffer area), commercial office park development (with natural buffer uses for the site), or some combination of the two scenarios (2003: 2). Its recommendations included actual site sketches showing the location and size of the natural buffers and areas designated for commercial and light industrial use (2003: 3-5). Although these uses were consistent with the existing zoning for the site and required no amendments to municipal ordinances to accommodate them, the committee's recommendations were adopted by the Hagerstown's City Council for inclusion in the city's Comprehensive Plan (2003: 18; Dukes 2005; Hagerstown Minutes 2004: 8). The committee recognized that additional information about the site could affect "types of appropriate land uses allowed at the site in the future" and urged EPA to "continue to work closely with the City of Hagerstown and community residents in the future to address community concerns and work with the community to clean up the Central Chemical site and return the site to successful reuse" (Hagerstown Land Use 2003: 39).

Community Views

Another related portal through which market considerations and other values important to the community may enter site deliberations along with environmental protection is the requirement that EPA solicit community views and consider the acceptability of the Agency's preferred remedy to the state and the community in selecting a remedy. The state and the local community typically have strong concerns about the environmental risks at a given site, but they also may have concerns about other issues. The state may be concerned about operation and maintenance costs that it will have to shoulder. The community may be concerned about the effects of the remedy and future uses of the site on jobs, property values, tax revenues, quality of life, the identity of the neighborhood, as well as the environmental justice implications of such decisions. Under EPA regulations, acceptability of a remedy to the state and community is a consideration that comes relatively late in the process. As "modifying criteria," the acceptability of the remedy assumes, at least nominally, a less central role in EPA's deliberations than the "threshold" protectiveness criteria or even the "primary balancing criteria" such as effectiveness and implementability. Nevertheless, like consideration of land use, consultation with the state and local community on remedy selection provides a vehicle for a broader range of concerns to enter the process. In most cases, the local community is the primary if not the sole bearer of the environmental risks posted by the site, and its views will therefore bear importantly on the relative environmental value of various cleanup scenarios. The local community also stands to reap a substantial portion of the non-environmental benefits of cleanup, including the benefits that flow from reuse of the site, and may also be in the best position to assess those benefits. The next section explores collaborative processes designed to develop and order community preferences relevant to site management decisions in both the public and private sectors.

Deliberative Process

The Superfund statute contemplates a decision process for remedy selection that is deliberative rather than technocratic. A classic bureaucratic or technocratic decision model might be appropriate if the statute provided rules of decision that could be applied mechanically. But, as we have seen, the statute and EPA regulations identify general factors to be balanced, provide some soft signals about how they are to be weighed, and create avenues for still other values to influence the process. An EPA decision-maker strikes the final balance, but she does so with limited guidance from the statute as to which outcome should be favored.

To help guide EPA's deliberations, the Superfund statute and EPA regulations and guidance provide for "community involvement," which is EPA's term for its process of informing the affected community about the site and considering its advice (EPA 2002b). The statute and regulations require the Agency to consult with the community during the RI/FS process; hold a public meeting on its proposed remedial plan for the site; provide an opportunity for public comment before the remedy is selected; and consider the acceptability of the plan to the state and the community (40 C.F.R. § 300.430(c)). In response to public reaction to the proposed plan, EPA is required to reassess "its initial determination that the preferred alternative provides the best balance of trade-offs, now factoring in any new information or points of view expressed" (40 C.F.R. § 300.430(f)(4)). EPA community involvement guidance encourages EPA site teams to go beyond the "letter of the law" by engaging the community early and seeking and considering its input throughout the process—from initial site assessment to post-cleanup monitoring and deletion from the National Priority List (Davies 2001; EPA 2002b: 23-28). At sites with high levels of interest, EPA also encourages establishment of Community Advisory Groups (CAGs), representing diverse community interests, to consult with EPA and state and local governments (EPA 1995a; EPA 2002b: 33-34).

Despite the commitment reflected in its guidance, commentators have been critical of the Agency's engagement of stakeholders.

> [C]urrent practice treats stakeholder participation as a constraint—i.e., potentially controversial alternatives are eliminated early. Little effort is devoted to maximizing stakeholder satisfaction; instead the final decision is something that no one objects too strenuously to. Ultimately, this process does little to serve the needs or interests of the people who must live with the consequences of an environmental decision (Linkov et al. 2004: 41).

Collaborative Stakeholder Processes

As a response to such criticisms, adaptive management supports a more widespread use and monitoring of collaborative stakeholder processes throughout the Superfund program. Three features of such processes are of particular importance within an adaptive management framework for Superfund. First, collaborative stakeholder processes are value-driven; they are dedicated not merely to reducing conflict over EPA decisions, an often cited purpose of the Agency's community outreach efforts, but more fundamentally to enhancing value to those primarily affected by such decisions. EPA remains the primary steward of certain values, for example, by assuring minimal protectiveness and husbanding the fiscal resources of the Superfund program, but acknowledges the community as the source of other values crucial to its decisions.

Second, these processes provide a forum for clarifying and ordering values that are typically not well-defined or prioritized in terms of preferred site outcomes or objectives. At Superfund sites, stakeholder preferences are likely initially to be unclear or misinformed, because of the unfamiliarity of the issues, including the technicalities of risk analysis and remedial selection and design, and uncertainties surrounding the sites (McClelland et al. 1990: 495; Gayer et al. 2000: 446). Thus, stakeholder preferences are amenable to shaping by a process of consideration (i.e., they are at least to some degree endogenous rather than exogenous to the decision process) (Freeman 1997: 53). In the sustained deliberative process contemplated by adaptive management, the public interest emerges through a focused interaction, in which consideration of technical and scientific information about the site and alternative scenarios combines with evaluation.

Third, these processes involve policy learning, which is the development and refinement of community preferences over time as uncertainties are resolved and more is understood about the site and its possible futures (Sabatier 1988). Policy learning requires that collaborative efforts commence at the earliest stages of a Superfund site inquiry and continue as long as decisions remain to be made. It also requires that stakeholders help frame working hypotheses about site conditions and alternatives and identify questions to be answered by activities at the site. Rather than soliciting one-time public reactions to data, policy learning contemplates an ongoing dialectic between technical and scientific information and values affecting policy choices.

Collaborative stakeholder processes embodying these features may be most effectively carried out through relatively small, continuous, representative groups, like the eighteen-member Land Use Committee that considered the future of the Central Chemical site in Hagerstown. The group should be large enough to provide balanced representation of diverse interests, including those of immediate neighbors of the site, facility owners, and other responsible parties; the local business and real estate communities; environmental and other community public interest groups; and local government (EPA 1995a: 7-8). At the same time, keeping the group as small as possible within this constraint reduces transaction costs, enhances the development of norms of reciprocity and trust, and increases the possibility that community members with different interests and values will find agreement (Olson 1965: 51-65; Cannon 2000: 408). The continuity of these groups, particularly with regard to stability in their leadership, is also important to their success as instruments of policy learning (EPA 1996: 5).

Community Advisory Groups (CAGs)

As described in a 1995 guidance document, EPA supports the use of Community Advisory Groups (CAGs) at sites where there is a high level of interest and concern about the site or where there are environmental justice concerns (EPA 1995a: 3; EPA 2002b: 33-34). Agency studies have found that such groups are more effective in clarifying concerns and resolving issues than public meetings; the same studies also show that a CAG can give the community more influence in site decisions (EPA 1996: vii). Despite the Agency's endorsement of CAGs, they have been used at very few Superfund sites; the Agency has only used CAGs at approximately 6 percent of eligible NPL sites (Leahy 2004). There may be several reasons for this. First, because of limitations on federal advisory committees imposed under the Federal Advisory Committee Act (FACA), EPA has concluded that it may not act directly to establish a CAG, although it may encourage a CAG's formation (5 U.S.C. app. §§ 1–14 (2000); EPA 1995a: 8; EPA 2002b: 33). Instead, the community—i.e., one or more local stakeholders—must establish the CAG. This requirement makes the formation of a CAG contingent to a significant degree on the community's self-organizing capabilities. Under FACA, EPA may not select a CAG's members, but EPA guidance (1995a: 8) requires that it certify the group's representativeness, and thus the Agency retains some leverage to ensure diversity and balance. Second, CAGs absorb EPA resources, including significant time of the EPA Community Involvement Coordinator (CIC); these are resources that may, in the judgment of EPA officials, be better employed elsewhere. Finally, because CAGs typically increase the influence of the community in the decision process, some EPA officials may be concerned that they will have correspondingly less control. This concern about control may be joined with concerns that if the CAG does not function well, it will impair rather than enhance the decision process.

Although formal collaborative structures will certainly not be warranted at every site, evidence that CAGs have the potential to work as adaptive management vehicles supports arguments for a greater effort to expand and perfect their use. Congress should consider amending the Superfund statute, FACA, or both to encourage the use of CAGs and to empower EPA to take a more direct role in establishing and supporting them. Even in the absence of statutory changes, EPA should ratchet up its commitment to CAGs, including not only technical and administrative support but also strategic use of its site decision-making authority to reduce the risks of failure inherent in collaborative undertakings.

Avoiding Failure

Two common modes of failure—capture and stalemate—are of special concern for CAGs and other collaborative processes at Superfund sites. Even if they include a representative range of stakeholder interests and values, collaborative groups like CAGs run the risk of capture: domination by sophisticated, well-organized interests that may unduly skew the deliberations in their favor. Capture is a risk in almost all policymaking settings, but some commentators have argued that it is more likely to occur in local, collaborative forums than in the context of centralized rule making (Karkkainen 2003: 961; *but see* Revesz 2001: 553) and thus poses a particular challenge for increased EPA reliance on CAGs. EPA can reduce this risk through its technical assistance grants (TAGs), which fund community groups (CAGs may qualify) to hire independent technical advisors to interpret information, or through the Agency's technical outreach services for communities (TOSC) program, which provides independent technical advice through EPA research centers (EPA 1998: 21-25). This expert assistance can counter the disadvantage that lay citizens may experience vis-à-vis more sophisticated or well-resourced players (EPA 2002b: 31; EPA 1996: 4-5). Although EPA (1998: 21-25) encourages community groups to apply for these resources, only about half of the CAGs at NPL sites have had technical assistance through TAGS or TOSC.

The Agency can also ensure, as it must by law, that its final remedy decision is consistent with CERCLA's criteria, including protectiveness (42 U.S.C. § 9621(c)), in order to provide a base level of protection against overreaching by interests that may be better organized, better informed, or otherwise advantaged in the process. More particularly, EPA can signal to the stakeholders that, in considering acceptability to the community among the statutory criteria, the Agency will not defer to CAG recommendations that do not reflect a reasonable accommodation of the range of local stakeholder interests and values.

Collaboration may also be undermined by strategic behavior among stakeholders, such as "stonewalling, strategic bargaining, dilatory tactics, and other forms of unilaterally imposed transaction costs, tending inevitably toward stalemate or least-common-denominator outcomes" (Karkkainen 2003: 964). Although these behaviors may occur in any negotiation, they may be particularly likely in situations where the consequences of failure to agree are unclear, and thus the parties have little incentive to cooperate (966), as may be the case under the relatively open-ended statutory criteria for remedy selection under Superfund. EPA can create incentives to bargain by giving notice to the parties of the remedy that it is considering adopting in the absence of a recommendation, and thus giving the stakeholders a distinct point to bargain around (Karkkainen 2003: 965-970; cf. EPA 2002b: 31). This signaling will occur formally with the Agency's issuance of its "proposed remedy" but could also occur earlier in the process as EPA discusses remedial options informally with interested parties.

Working through CAGs requires EPA to balance several roles: enabler of the collaborative process, setter of boundaries and provider of incentives to ensure that the process is effective and fair, and final arbiter of the remedy (John 1999: 19). These roles are in tension, and managing them effectively may be among the greatest challenges of a more decentralized model for Superfund. As with technical and scientific issues, the design and implementation of collaborative mechanisms at each site are subject to contingencies, and thus are properly the subject of "learning by doing" within an adaptive management frame.

One might question whether EPA can be trusted to carry out these roles as an honest broker. For example, one might be concerned that the Agency would be reluctant to truly empower local stakeholders or give appropriate deference to their recommendations out of an institutional reluctance to limit its own policy prerogatives by sharing "turf." In enacting Superfund, Congress deemed it in the national interest that the Agency be solely entrusted with making the final remedy decision, but this arrangement carries with it the risk of self-serving central bureaucratic behavior that could discourage or distort local stakeholder processes at some cost to the broader public interest. Limiting this risk are the institutional benefits that EPA stands to gain from producing value-enhancing results at the community level. The Agency's popular brownfields program, which began as an agency initiative to facilitate the reuse of contaminated sites not dealt with under Superfund (EPA 2005a), evidences the Agency's recognition of this, as does the Agency's recently increased emphasis on community involvement and site reuse under Superfund (Davies 2001; Cook 2002). Finding out what affected communities want and helping to give it to them, in appropriate measure, can generate political capital for the Agency in the White House and Congress and among the general public—capital that the Agency can use to preserve its authorities and protect its budget. Thus the Agency's most fundamental institutional interests, properly understood and implemented, are likely to be congruent with well-supported, empowered, and balanced stakeholder collaborations.

Quantitative and Qualitative Techniques

Although they do not provide a substitute for a deliberative process, quantitative techniques such as cost-benefit analysis and multi-criteria decision analysis (MCDA) may be useful as deliberative aids. One limitation of these approaches is the assumption that values or preferences are fixed and pre-existing and thus can simply be aggregated (cost-benefit analysis) or otherwise systematically sorted, ranked, and applied (MCDA) to indicate the desired or optimal result (Prichard et al. 2000: 38-39; Norton & Steinemann 2001: 478-479). Some forms of MCDA, however, retain the flexibility to "allow stakeholders to 'change their minds'" by adjusting the relative weightings given to selection criteria or "by introducing new criteria or alternatives at any time during the analysis" (Linkov et al. 2004: 10). MCDA has a further advantage over traditional cost-benefit analysis in its ability to account systematically for preferences or values that are not related to any economic use or value (Linkov et al. 2004: 1). And it has been applied with some success in collaborative settings "as a framework that permits stakeholders to structure their thoughts about pros and cons of different remedial and environmental management options" (Linkov et al. 2004: 25; NRC 2003: 106-107). The disciplined thinking involved in these analytical techniques may help stakeholders counter the effects of "cognitive problems" that have been identified as affecting environmental decision-making. For example, stakeholders may have difficulty accurately assessing the range and probability of possible outcomes in a context of uncertainty (Sunstein 2002: 9, 26-27). Systematic analysis and quantification of contingencies, even though subject to uncertainties themselves, can sharpen the judgment of stakeholders about how to factor uncertainty into their deliberations. These methodologies may also improve the deliberative process by limiting the ability of interest groups to exploit distorted perceptions of risks and probabilities to their advantage.

Qualitative techniques may also assist the deliberative process involved in Superfund cleanups. Examples include historical reviews of the site and its environs, local cultural studies, and architectural designs that visualize alternative uses as physical continuations of the site (Center of Expertise for Superfund Site Recycling 2005). These tools can open up value-enhancing possibilities that site owners and developers, as well as local, state, and federal governments, each thinking separately in their traditional ways, might not explore (Lee 1999); stakeholders can then jointly assess these possibilities. A scenario that emerges successfully from the process may not only coalesce support for a particular remedial option but also coordinate successive actions among governmental entities and across the public and private sectors.

FLEXIBILITY, UNCERTAINTY, AND CHANGE: DOING AND LEARNING

The Superfund process is extended in time and made up of myriad information-gathering activities and decisions. These activities and decisions include the initial identification and scoring of the site; listing on the National Priority List (NPL); the Remedial Investigation/Feasibility Study (RI/FS) (including the reasonably anticipated future land use determination); remedy selection, implementation, and evaluation; possible remedy revision; deletion from NPL; and post-remedial five-year review for the roughly 60 percent of "construction complete" sites where some residual waste remains on site. This process is lengthy. EPA has estimated the average time from proposal for listing on the NPL to completion of the remedial action at approximately eight years, but a recent study by Resources for the Future calculates the average instead at over eleven years (Probst et al. 2001: 47-52; E^2 Inc. 2005: 3–37 to 3–38.). Actually achieving final cleanup goals may take much longer in some cases; at sites with long-term remedial actions such as bioremediation and soil vapor extraction final cleanup can take twenty years or more (Probst et al. 2001: 49). For the roughly 60 percent of sites where waste remains on site after completion of the remedy, monitoring and review are mandated for as long as contamination remains above a level that allows "unlimited use and unrestricted exposure" (EPA 2001a). For some Department of Energy sites involving radioactive contaminants, the projected period of agency involvement extends for thousands of years (NRC et al. 2003: 8-10).

Adaptive management leads us to think of this process as a series of interventions over time, with the aim of ensuring that each intervention is informed by current information, including information about what occurred in response to previous interventions. Adaptive management anticipates that decisions will leave maximum flexibility for later adjustments and that they will be revisited and revised, if appropriate, in light of new information. It also anticipates effective coordination of site decisions made by the public and private sectors and by multiple levels of government. Achieving these management characteristics in Superfund will depend on having institutions that (1) provide for adequate monitoring and feedback mechanisms (information flow); (2) do not foreclose options unnecessarily while proceeding with the tasks of decision and implementation (flexibility); (3) enable revisiting and adjusting prior decisions as warranted (self-criticality); and (4) integrate across sectors and jurisdictional scales (hierarchical linkages). The discussion that follows explores how such institutions might work.

The Superfund process as currently defined demonstrates some ability to accommodate an adaptive approach, including required monitoring and modification of remedies to ensure protectiveness over time. Recent initiatives further demonstrate EPA's willingness to review and adjust prior remedial decisions and reuse determinations in light of new information or changed circumstances (EPA 2004c; EPA 2005g). However, Superfund's learning process remains largely reactive and without systematic articulation or justification. These shortcomings are particularly evident in the technically and institutionally complex co-evolution of the remedy and reuse plans for the site—a dance that, as we have observed, is essential to a value-enhancing disposition of the site.

Adaptive Management in Superfund Site Remediation

From Site Study to Remedy Completion

As mentioned above, it takes more than eleven years on average for a Superfund site to move from proposed listing on the NPL to completion of the remedy. This average is likely to grow even longer in the future. To meet the Agency's "construction complete" goals, EPA regional managers have focused on sites for which remedy construction could be completed quickly; as a result, many of the sites remaining "require more complex, lengthy, and expensive cleanups" (Probst & Sherman 2004: 3). A significant portion of the work remaining is concentrated at "mega sites"—sites whose cleanup costs exceed fifty million dollars, and whose high cost and technical, scientific, and institutional complexity can greatly extend the time to remedy completion (NACEPT 2004: 69-71). Current and anticipated funding constraints will almost certainly extend the average time for cleanup even further (Dunne 2004: 7). The process defined by EPA regulations and related guidance provides multiple decision opportunities during this extended period. This process therefore allows adjustments to be made in response to new or evolving information, and thus for the application of adaptive management principles.

During the first phase, EPA conducts the remedial investigation/feasibility study (RI/FS) and selects, designs, and implements a remedy. The remedial investigation (RI) characterizes the site, conducting field studies and a baseline risk assessment, and sets protectiveness goals that are used to develop remedial alternatives and to measure the efficacy of those alternatives. EPA regulations recognize that "estimates of actual or potential exposures and associated impacts on human and environmental receptors may be refined throughout the phases of the RI as new information is obtained" and therefore that these goals may change during the remedial investigation (see 40 C.F.R. § 300.430(d)). In the baseline risk assessment, discussed above, the reasonably anticipated future land use is determined and used to set cleanup objectives (EPA 1989). By implication, assumptions about future land use, along with other elements of EPA's risk assessment, will be reviewed and refined in light of "new information" prior to the selection of a remedy.

The feasibility study (FS), which is developed in coordination with the RI, defines and assesses (practicable and cost-effective) remedial alternatives to meet cleanup objectives for the site. EPA regulations require that EPA's assessment of alternatives take into account uncertainties affecting the success and long-term effectiveness of the remedy. Information gleaned during the RI/FS may help to reduce them or in managing them effectively over time.

In the remedial decision, EPA selects the remedy that will be undertaken at the site and issues a record-of-decision (ROD). Issuance of the ROD is followed by remedial design and implementation (RD/RA). During this post-ROD phase, conditions at the site may change and new information will certainly emerge during the construction and evaluation of the remedy, including sampling and analysis to determine whether cleanup levels have been achieved by the remedy (40 C.F.R. § 300.435(b)). Negotiations with responsible parties to carry out the cleanup with private funds rather than with public funds may occur during the RD/RA phase. The outcome of these negotiations may affect cleanup plans. In provisions that recognize that there may be changes in the ROD between its issuance and final implementation, the NCP provides a mechanism by which new information and developments bearing on the remedy can be considered and acted upon during the RD/RA (40 C.F.R. § 300.435(c)(2)). ROD amendments provide an important adaptive management tool, both during and after the RD/RA, and pursuant to a program reform begun in 1996 to update remedy decisions, there is some evidence that this tool is being used effectively in response to new information generated during the remedial design process (Luftig & Breen 1996; EPA 2003).

EPA regulations do not expressly address how EPA is to manage uncertainty or new information affecting the reasonably anticipated future land use during the RI/FS. EPA guidance states that "where the future land use is relatively certain, the remedial action objective generally should reflect this land use." If uncertainty surrounds the reasonably anticipated future land use, "a range of the reasonably likely future land uses should be considered" (Land Use Directive 1995: 8). Each reasonably possible land use may be consistent with some remedial alternatives, including engineering measures and institutional controls, but not with others (Ferguson 2002: 17, 21-24). Thus, the guidance directs that "[t]hese likely future land uses can be reflected by developing a range of remedial alternatives that will achieve different land use potentials" (Land Use Directive 1995: 8). By developing this range, the Agency retains flexibility to respond to new information on future land use at least through remedy selection.

Similarly, EPA regulations make no specific mention of future land use in the RD/RA phase. EPA guidance states that the remedy selection process includes determination of "[l]and uses that will be available following completion of the remedial action" and planning of site activities that are "consistent with the reasonably anticipated future land use" (Land Use Directive 1995: 2). Where reuse plans are well developed at the time of remedy selection, the remedy can be tailored in its design and implementation to ensure both protectiveness and the realization of the reuse plans. In some cases, where site preparation requirements for future development may exceed what is necessary to achieve a protective remedy, arrangements can be worked out to accommodate those requirements prior to implementing the remedy. For example, at the Raymark site in Stratford, Connecticut, the remedy chosen was an engineered containment system. The prospective developer paid for dynamic compaction and the installation of pilings during the construction of the containment system in order to support future building on the site (EPA 2002a: 7-8). Thus, by reducing the total costs of remedy implementation and site preparation, the net value of the cleanup and redevelopment of the site can be increased.

Reuse plans are often not well developed at the time of remedy selection, in which case the Agency may make protective remedy decisions that anticipate the most likely category or categories of redevelopment and preserve maximum flexibility for future adjustments. Under its general remedy revision authority, discussed above, the Agency may modify its remedy to accommodate proposals that emerge during the RD/RA phase. For example, the remedy for the Rentokil, Inc. site, a former wood treating plant in Henrico County, Virginia, provided for removing wood treating equipment and some contaminated sediments and for building control structures to reduce further migration of contaminants into a creek (EPA 2002a: 41). During implementation, the remedy was revised to provide for redevelopment, allowing building foundations to be incorporated into the cover and other structures necessary for construction and consistent with long-term maintenance of the remedy. The Rentokil site illustrates the importance, as part of an adaptive management approach, of retaining flexibility to respond to new information on land use preferences as well as other value dimensions of the site even after the remedy has been selected and is being implemented.

Because future land use is so heavily dependent on decision-makers other than EPA—the site owner, potential developers, local land use authorities, and state officials—and because future land use and the evaluation and implementation of remedial measures are so closely related, focused stakeholder processes of the sort discussed above provide a critical adaptive management tool. These processes can provide information on evolving stakeholder preferences relating to site planning and management over the course of the RI/FS and the DR/RA and even beyond. Moreover, collaborative stakeholder processes can help integrate perspectives across multiple scales within the decisional hierarchy.

Site Reviews

The period after completion of the remedy presents perhaps an even broader field for adaptive management, and yet the potential for adaptive management in this field is not well developed in the EPA regulations and guidance, particularly regarding land use. For sites on which hazardous substances remain, EPA conducts post-remedial inspections and reviews. For remedies adopted after the Superfund amendments of 1986, the statute requires that the Agency conduct an in-depth review of the effectiveness of the remedy at such sites every five years after initiation of the remedial action (42 U.S.C. § 9621(c) (2000); EPA 2001a: 1-3 to 1-4). This review produces (1) an assessment of whether the remedy is protecting human health and the environment and (2) recommendations for actions that need to be taken to ensure continued effectiveness (if the remedy has performed adequately to date) or to restore protectiveness (if it has not). The Agency's review includes community notification, a site inspection and interviews, a review of data from site monitoring, and "any other information [that has] come to light that could call into question the protectiveness of the remedy" (EPA 2001a: 3-7). The Agency may require additional sampling and collection of other data as necessary to decide whether the remedy is functioning adequately (2001a: 4-12 to 4-13). Five-year reviews may be discontinued only when the Agency determines that contaminant levels on site are below levels that allow for unlimited use and unrestricted exposure (2001a: 1-4).

The five-year review plays a central role in the long-term stewardship of NPL sites, but EPA could significantly improve its usefulness as an adaptive management tool. The five-year review gives the agency both the occasion and the information on which to act, if the protectiveness of the remedy is in question (NRC 2003: 303). The review is crafted particularly to ensure the continued effectiveness of engineering measures designed to contain remaining on-site contamination (such as caps) and institutional controls designed to limit human exposure (such as land use restrictions). Recently EPA took steps to enhance its "ability to gather, manage and evaluate" information on institutional controls through the five-year review (EPA 2004b: 4).

However, EPA's five-year review guidance does not direct the Agency to inquire into dimensions of the remedy/reuse other than protectiveness or to take action if it is apparent that the remedy is not functioning well along these dimensions. For example, the guidance does not require investigation of whether the remedy remains cost-effective, that is, whether changed conditions at the site or new technological information indicate that a modified remedy could be operated or maintained at lower cost (EPA 2001a: 4-4). Although EPA's reform initiative to update remedies extends by its terms to all RODs, whether construction of the remedy is complete or not, most of the remedy updates have occurred in the design phase (EPA 2003: 3). Requiring focused examination of this issue during the five-year review could provide additional information for post-completion sites.

The available guidance also affords only limited consideration of site reuse. It does direct EPA to consider "changes in land use" as part of its review, but such changes are only relevant to the issue of whether unanticipated exposures have undermined the protectiveness of the remedy

(EPA 2001a: 4-5). There is no consideration of whether the reuse option anticipated by the remedy has been carried out, whether the current use represents a productive use of the site or otherwise accords with the wishes of the community, or whether alternative uses have materialized that promise a more locally acceptable use of the site without compromising the remedy's protectiveness. Recently the Agency has expressed interest in facilitating reuse at post-completion sites. Late in 2004, EPA announced a "Return to Use" initiative, "designed to remove barriers to reuse that are not necessary for the protection of human health, the environment, or the remedy at those sites where remedies are already in place" (EPA 2005g). The eleven demonstration projects selected for that initiative are a beginning, but broader program guidance will be necessary to secure review of all post-completion sites, including the more than four hundred such sites that are not in productive use (EPA 2005h). The Agency has stated its intent to issue guidance to integrate "consideration of reuse throughout the response cycle," including "long-term stewardship" (Cook 2002: 2). In framing that guidance, the Agency could make good use of the five-year review.

In sum, the five-year review is ideally suited to carry out adaptive management of sites after remedy completion but, as it is currently structured, adaptive management principles and processes are applied only to some of the factors that are relevant to enhancing site value. Dimensions related to non-environmental values that may be important to the community are not integrated, and the review therefore wastes an opportunity for EPA, in consultation with the state, local officials, community groups, and business interests, to further the public interest as it is broadly understood.

It remains open at any time for an interested party to seek modification of a remedy to accommodate a different, more beneficial use (40 C.F.R. § 300.435(c)(2); EPA 2004d: i). A town may propose remedy enhancements that would allow use of a site as a park, or a developer may propose changes that would allow more intensive (i.e., higher exposure) uses of the site. If the proponent will fund the enhancements and can persuade the Agency that the amended remedy will be protective and consistent with the other statutory criteria, EPA may approve it. That less than one-third of construction complete sites have been returned to productive use suggests that there is significant unrealized potential at these sites (EPA 2005k). Because of the stigma attached to hazardous waste sites, however, underutilized Superfund sites may not receive the attention either from private development interests or public entities that other properties might (McCluskey & Rausser 2003: 276; Mundy 1992: 7). The five-year review offers a strategic opportunity for EPA to re-engage the community on the issue of site utilization as well as on the issue of protectiveness, and to facilitate alternative uses of the site where it is determined that the land is being underutilized. The characteristics of Superfund sites, as among a relatively small group of the most contaminated sites in the nation, are not within the range of information or expertise typically possessed by local real estate markets or local governments. Niche entrepreneurs specializing in developing contaminated sites may help to supply that information and expertise to the affected community, but EPA remains an important and arguably more transparent additional source. Certainly where the Agency is required to maintain prolonged contact with the site, as under the five-year review provisions, it seems appropriate that its ongoing consultations with local private and public interests provide a forum for ensuring that the site is well-used as well as safe.

HIERARCHICAL LINKAGES: INTEGRATING ACROSS SCALES

Adaptive management attends to hierarchical linkages, in both natural and human systems. It calls on EPA and others who make decisions affecting Superfund sites to locate their understanding of the site's physical and biological resources in the larger physical and biological systems to which they belong. It also calls on decision-makers to understand their place within the institutional hierarchy that affects the site. Because Superfund sites—as distinct from other categories of land generally managed by private markets and local regulation—experience a substantial federal presence, the hierarchical considerations affecting these sites are both unusual and complex.

To understand how the hierarchical aspects of Superfund sites might best be addressed within an adaptive management framework, it is helpful to understand the possible theoretical rationales for federal involvement in these sites. Economists offer the subsidiarity or matching principle to determine the level of government at which regulatory decisions should be made: the decisions should be made at the smallest unit of government whose geographic scope includes all the significant costs and benefits of the regulation (Oates 1972: 31-38; Butler & Macey 1996: 23, 25). The matching principle supports federal regulation where a localized activity, such as site contamination, would have significant environmental or economic effects in other states. Physical interstate spillovers of the sort generally acknowledged to warrant federal intervention are not apparent at most Superfund sites, as the effects of soil and groundwater contamination tend to be geographically confined. The Eleventh Circuit, however, in *United States v. Olin Corp.*, 107 F.3d 1506 (11th Cir. 1997), held that Superfund fell within the federal Commerce Power, noting comments in the legislative history that the "improper disposal of hazardous waste threatened natural resource-dependent, interstate industries, such as commercial fishing" (107 F.3d at 1511 n.10). The court also cited Congressional findings that "accidents associated with purely intrastate, on-site disposal activities" adversely affected interstate commerce, and it concluded that "the regulation of intrastate, on-site waste disposal constitutes an appropriate element of Congress' broader scheme to protect interstate commerce and industries thereof from pollution" (107 F.3d at 1511 n.11). Nevertheless, the interstate externalities argument for Superfund does not seem particularly strong compared to similar arguments for other federal environmental statutes, such as the Clean Air Act and Clean Water Act, in which interstate pollution problems figure much more prominently.

There are at least two other possible theoretical justifications for the federalization of programs for cleaning up seriously contaminated sites. First, the contamination of many of these sites is traceable to the business activities of large national or multinational corporations. Given the size of these corporations, their substantial economic leverage within the individual states, and the hefty costs of cleanup, one might argue that a federal liability system is necessary in order to prevent a race to the bottom in cleanup programs among states, resulting in too little cleanup. This possible destructive interstate competition provides an independent justification for a federal scheme, and indeed the race to the bottom was cited by legislators and courts to justify the centralized decision structure of Superfund. However, Richard Revesz and others have criticized the race-to-the-bottom rationale in environmental regulation as an insufficient theoretical basis for the federalization of environmental programs (Revesz 1992: 1233-44; Oates 1997: 1325-27; *but see* Esty 1996: 627-38). There may be competition for economic development, Revesz argues, in which states are forced to balance their desire for jobs with their preference for a clean environment, but that competition is not necessarily a race to the bottom; indeed, it may be welfare-enhancing (1992: 1242).

Second, it might be argued that the states lack the capability (the scientific, technical, or legal sophistication) to effectively deal with the largest and riskiest contaminated sites. Only the federal government, with its advantages of scale, can marshal the requisite expertise to manage these sites effectively, and thus the federal regime is justified (Sarnoff 1997: 251-57). This argument may be partially offset by the geographic heterogeneity of contaminated sites, where "on-the-ground knowledge is of central importance, and the diversity of circumstances is salient" (Esty 617). Moreover, even granting superior technical, scientific, and legal capabilities to federal officials, this rationale does not necessarily support placing sole decision-making authority at NPL sites in the hands of federal officials, as Superfund does: the federal government might simply make its expertise available to state or local decision-makers.

The federal presence at NPL sites and other contaminated sites warranting emergency response serves important functions, but the possible theoretical justifications for that presence do not provide overwhelming support for federal hegemony. The interests and capabilities of states, localities, and private parties in the management of Superfund sites justify a substantial and ongoing role for them in site-related decision-making. Typically, the benefits of cleaning up and redeveloping a Superfund site are realized predominantly within the state and, indeed, within the local jurisdiction in which the site is located. A significant portion of the costs of cleanup and reuse are also likely to be felt within the state and the locality. Even if federal funds are used for cleanup, spreading most of the remedial costs nationally, the state remains obligated for a share of those costs and for long-term operation and maintenance costs as well. Moreover, the land use aspects of Superfund sites fall within the traditional purview of state and local regulation. Accordingly, adaptive management in Superfund site management suggests that EPA invest heavily in processes to elicit the preferences of state and local stakeholders throughout its involvement at the site—with particular emphasis on the community where the impacts of site activities will be concentrated—and to facilitate integration of the results into federal, state, and local decisions affecting the site.

ADAPTIVE MANAGEMENT OF THE SUPERFUND SITE INVENTORY

Adaptive management principles are also applicable to EPA's management of the Superfund site portfolio as a whole. Program-level issues that might benefit from continuous learning include what remedies work best in particular types of sites or with particular types of contaminants; what remedies work best with specific types of land use; what community involvement techniques are most effective in eliciting useful and reliable information about community preferences; how best to integrate decisions across private and public sectors and across federal, state, and local jurisdictions; and what the relevant contingencies are and how best to address them.

Institutional learning has been going on since the program began, but much of it has been episodic and reactive. Since 1989, EPA has conducted at least three comprehensive agency-level studies of its management of the program (EPA 2004c: 17-18). In addition, EPA's Inspector General, the General Accounting Office (now known as the Government Accountability Office), and non-governmental groups have reviewed the program with sporadic zeal (EPA 2004c: 18-19). These studies typically begin with a set of problems, concerns, or allegations, then assemble data relevant to the issues, and finally conclude with a set of findings and recommendations for addressing the problems, concerns, or allegations. They represent a rough form of trial-and-error learning, which tends to be driven by perceived program failures or impending crises rather than conscious policy experimentation or, more modestly, continuous monitoring of key program indicators and corresponding program adjustments.

EPA could improve Superfund policy learning program-wide by acknowledging and addressing complexity and uncertainty in program implementation; framing policies to test hypotheses about how the program might work better and carefully monitoring their implementation; and, even more fundamentally, systematically monitoring and recording experience at sites as a basis for ongoing review and adjustment of national policies. The last of these—generating and recording site information—is crucial for continuous learning, and it is an area of particular difficulty for Superfund. Despite the existence of various Superfund databases, Katherine Probst and Diane Sherman found that "it is difficult to obtain reliable information on key attributes for [NPL] sites" without talking to regional staff directly involved with the site (2004: 7). Well-developed case studies of Superfund site decision-making are scarce. Agency documents, such as the RI/FS and the ROD, are prepared as part of the administrative record at each site, but do not record all the steps leading to the decision reached—for example, the nature and success of collaborative efforts, the scientific and technical uncertainties addressed, or the lessons learned. There are also few well-developed accounts of the post-ROD process, including remedy review and reuse decisions. The absence of such accounts makes it very difficult to determine, among other things, frequently occurring contingencies and the most effective responses or moderating measures for those contingencies for the Superfund universe as a whole.

Probst and Sherman recommend that the Agency develop a core set of data for each site that includes "important measures of progress as well as key site attributes" and "that meets the needs of the full panoply of stakeholders" (2004: 8). They see a consistent, well-maintained site monitoring and reporting system not only as a means of improving management of particular sites but also as a source of aggregated data for improving overall program efficiency and effectiveness (2004: 9). Similarly, the recent report of the Superfund Subcommittee of EPA's National Advisory Council for Environmental Policy and Technology (NACEPT) recommended that EPA "develop and implement a system to ensure clear, transparent dissemination of a core set of data for all NPL sites and Superfund program activities" (NACEPT 2004: 5-88). Properly detailed, these data could provide the basis for more effective integration of diverse stakeholder perspectives within Superfund's complex hierarchical setting.

These criticisms notwithstanding, Superfund has established practices that facilitate learning program-wide. These practices include using pilot projects to test new policies or strategies before widespread implementation (EPA 2005d); facilitating the testing of new cleanup technologies and collecting and disseminating information about those technologies (EPA 2005j, 2005b, 2004a: v); and tracking the success of program reforms and initiatives (EPA

2005f). Recently Superfund has even re-energized its effort to collect and disseminate timely and complete information on experience at individual sites (EPA 2005i). These practices provide a basis for further application of adaptive management principles to the program as a whole.

ISSUES AND CONCERNS

Environmentalist Concerns

Environmentalists might be troubled by the application of adaptive management advanced here, particularly the assumption that most sites will require long-term attention. Long-term management is only necessary, they might argue, where cleanups are less than complete. Doing cleanups "right" the first time—that is, choosing and implementing remedies that reflect the statutory preference for permanence and avoid the need for engineering and institutional controls—minimizes the need for long-term care and the uncertainties associated with such care. Markets and/or local government officials then have maximum flexibility to determine reuse without need of further involvement by EPA. Thus, rather than accommodating the current practice of leaving waste on site, reforms should focus on doing cleanups right.

It is certainly the case that long-term stewardship entails environmental risks and other costs that would not be incurred if every site were left in pristine condition at the conclusion of the remedy. But these stewardship costs may be justified as providing a more beneficial balance between protection and other values over time, including controlling the costs of initial cleanup; indeed, under current conditions of limited availability of funds, controlling remedial costs at individual sites can assure that more sites receive protection (NRC 2003: 38-42). Moreover, at sites where treatment or removal of all contaminants is simply not achievable, extended management must be provided in any event to assure continued protectiveness of the remedy. Because of the challenges presented to local decision-makers by the special characteristics of Superfund sites, expanding the federal management process to include ongoing consideration of land use potential and community wishes may be well worth the costs. Among other things, facilitating the productive use of sites may help secure management consistent with long-term public health protection.

Agency Concerns

One might expect the EPA to have policy and legal concerns about the expansion of its long-term responsibilities at sites. Particularly with Superfund under serious resource constraints (Hennelly 2003: 30), the Agency may see itself as having limited capacity for conducting more intensive, community-based inquiries into remedy/reuse options or for expanding its long-term stewardship obligations, such as by broadening consideration of land use in its five-year reviews.

The answer to this concern is similar to the answer to environmentalists. If one of the Agency's goals is to use public funds to maximize public benefit, then the program adjustments suggested here should be considered against alternative uses of agency resources. With limited funds, the Agency must triage among sites and among activities relating to a specific site. The argument is, at least with respect to specific sites, that greater public good is achievable through relatively inexpensive process changes, such as greater use of CAGs and expanding the five-year review process. The Agency must decide whether that argument is persuasive, and it must further decide whether the benefits of expending additional resources in site management justifies the potential impact on cleanup at other sites in the queue.

The Agency might also be concerned about perceptions that the adaptive management model suggested here, with its emphasis on the disposition of the site as a whole, including its future use, would convert Superfund into a federal land use program. By conditioning the uses to which land can be put and the terms on which those uses can be carried out, however, Superfund unavoidably intrudes on the process of land management normally carried out between private markets and local officials. Given the fact of that intrusion, it would seem incumbent upon the Agency to facilitate, to the extent it is able, a disposition of sites that reflects not only the market realities but also the citizen preferences of the affected locality.

Finally, the Agency (and others) might be concerned that EPA lacks the legal authority to implement the chapter's recommended changes, particularly the suggestion of continuing involvement with land use issues at Superfund sites. This chapter has suggested that this involvement is warranted, where contaminants remain on site, by provisions of CERCLA for selecting and maintaining remedies at NPL sites. But this involvement may be approaching the limits of the Agency's statutory authority and will have to be tailored to be consistent with that authority or additional legislative authorization obtained.

Responsible Party Concerns

Responsible parties may have reservations about the management approach suggested here to the extent that, to accommodate new learning, it would encourage ongoing attention to the uses of sites and to potential remedy changes. In particular, responsible parties that are also present site owners may resent the sustained intrusion of these considerations, threatening to turn a federal cleanup program into a federal land use program. It is not clear, however, that this approach would place additional constraints on the prerogatives of Superfund site owners. In its five-year review, EPA will determine either that a remedy continues to be protective or that it does not. The remedy may not be protective for any one of a number of reasons: the contamination may be more extensive than originally understood, the remedy may not be operating as effectively as projected, or the land use on which the remedy was predicated may have changed, leading to higher exposure than was contemplated by the remedy. In any of these circumstances the responsible parties, including any that are also site owners, may be liable for additional remedial actions necessary to assure protectiveness. This has always been the case under EPA's interpretation of the statute.

Under the approach suggested here, EPA, in consultation with the site owner and the community, would also use the five-year review to assess the current land use. Assuming the remedy is protective under the current use, however, EPA would have no authority based on its review to order a different land use; this remains a decision left to the owner and local land use authorities. An owner might seek to upgrade a remedy to accommodate a new use, but under EPA's legal interpretation, such an upgrade could not be compelled by the statute and would not be chargeable to the responsible parties unless undertaken by them voluntarily (EPA 2002a: 5). Thus, there would seem to be little or no additional cleanup liability risks for responsible parties in the extended attention to land use suggested here.

Imposing the costs of upgrading an otherwise protective remedy on the party or parties benefiting from the upgrade, rather than on the originally responsible party, tests the efficiency of a use that requires more intensive cleanup. A site owner will proceed with a new use only if the increase in the value of the property resulting from the new use outweighs the cost of upgrading the remedy and any other costs of development. If the owner is a public entity, presumably it will proceed with the new use only if the public benefits flowing from the new use outweigh the upgrade costs.

An important concern for many Superfund site owners is limiting liability that may be triggered by future actions on site. Owners of a number of large Superfund sites refuse to consider selling them or otherwise making them available for use by others, such as by lease, due to liability concerns (Quarles 2003). Their concerns include the fear not only that the actions or omissions of third parties on the site might occasion the need for additional remedial work (e.g., due to failure to maintain a cap or adhere to site use restrictions), but also and perhaps even more significantly, that third party access could expose the company to toxic tort litigation. These concerns are compounded by questions about the ability of institutional controls such as covenants or local land use regulations to ensure compliance over the long term (Strasser & Breetz 2003: 33-35). EPA has recently acted (2004b) to improve monitoring and enforcement of institutional controls, but adoption of a uniform environmental covenants act among the states could also help address this problem by binding subsequent owners to maintain the remedy and to take other precautions to minimize risks at the site (UECA 2003). Such legislation may prove attractive to disparate interests—property owners concerned about future liability, environmentalists concerned about the long-term integrity of the remedy, and developers desiring an increase in the number of properties available for development.

CONCLUSION

The broad transition that this chapter encourages would itself represent an adaptive response at the program level, as Superfund continues to mature from its early crisis response mode to long-term site management and as it responds to signals from the political system favoring attention to the values associated with reuse as well as reduction of environmental risk. The Agency can facilitate this transition by acknowledging its broader policy horizon, using policy as a tool for experimentation, strengthening program monitoring and reporting, and maintaining flexibility and openness to regular policy review and adjustment. Expanding Superfund's capacity for learning by doing will be critical as the Agency deals with new challenges such as limited funding, megasites, and terrorist threats and as it seeks more generally to define the continued relevance of Superfund to evolving societal needs and concerns.

REFERENCES

Butler, Henry N., and Jonathan Macey. 1996. "Externalities and the Matching Principle: The Case for Reallocating Environmental Regulatory Authority." *Yale Law and Policy Review* 14: 23-66.

Cannon, Jonathan Z. 2000. "Choices and Institutions in Watershed Management." *William and Mary Law and Policy Review* 25: 379-428.

Center of Expertise for Superfund Site Recycling. 2005. *Conference Information*, April 2004 University of Virginia conference on "Revitalizing Land and Restoring Communities," including presentations by Daniel Bluestone (local culture), Julie Bargmann (site design), and Niall G. Kirkwood (history). http://www.virginia.edu/superfund/conference.html (accessed October 6, 2005).

Cook, Michael B. 2002. Memorandum from Michael B. Cook, Director, Office of Emergency and Remedial Response, U.S. EPA, to Superfund National Program Managers Regions 1–10 & OERR Center Directors and Process Managers. *Reuse Considerations During CERCLA Response Actions* (October 10, 2002). On file with author.

Cook, Michael B. 2005a. *Superfund Site Progress Profiles—Status Report* (February 4, 2005). On file with author.

Cook, Michael B. 2005b. Telephone interview with Michael Cook, Director, Office of Emergency and Remedial Response, U.S. EPA, Washington, DC (March 30, 2005).

Davies, Elaine F. 2001. *Memorandum to Superfund National Policy Managers, Regions 1–10* (October 12, 2001). http://www.epa.gov/superfund/resources/early.pdf

Dukes, E. Franklin. 2005. E-mail from E. Franklin Dukes, Director, Institute for Environmental Negotiation, Charlottesville, Virginia (October 19, 2005).

Dunne, Thomas. 2004. Remarks by Thomas Dunne, Acting Assistant Administrator, Office of Solid Waste and Emergency Response, EPA, at the Superfund Seminar, Charlottesville, Virginia (December 2, 2004).
http://www.epa.gov/swerrims/docs/2004_1202_dunne_sf_speech.pdf

E^2 Inc. 2005. *Superfund Benefits Analysis* (partial review draft; January 28, 2005). http://www.epa.gov/superfund/news/benefits.pdf

Environmental Protection Agency, United States. *See* EPA.

EPA. 1989. *Risk Assessment Guidance for Superfund, Human Health Evaluation Manual.* http://www.epa.gov/oswer/riskassessment/ragsa/index.htm

EPA. 1995a. *Guidance for Community Advisory Groups at Superfund Sites.* www.epa.gov/superfund/tools/cag/resource/guidance/caguide.pdf

EPA. 1995b. *Superfund Administrative Reform Fact Sheet* (May 25, 1995).

EPA. 1996. *Community Advisory Groups: Partners in Decisions at Hazardous Waste Sites Case Studies.*

EPA. 1998. *Community Advisory Group Toolkit.*
http://www.epa.gov/superfund/tools/cag/cagtlktc.pdf

EPA. 2001a. Office of Emergency and Remedial Response. *Comprehensive Five-Year Review Guidance.* http://www.epa.gov/superfund/resources/5year/guidance.pdf

EPA. 2001b. *Reusing Superfund Sites: Recreational Use of Land Above Hazardous Waste Containment Areas.*
http://www.epa.gov/superfund/programs/recycle/tools/recreuse.pdf

EPA. 2002a. *Reusing Superfund Sites: Commercial Use Where Waste is Left on Site.* www.epa.gov/superfund/programs/ recycle/c_reuse.pdf

EPA. 2002b. *Superfund Community Involvement Handbook.*
www.epa.gov/superfund/tools/cag/ci_handbook.pdf

EPA. 2003. *Updating Remedy Decisions at Select Superfund Sites: Biannual Summary Report FY 2000 and FY 2001.*
http://www.epa.gov/superfund/programs/reforms/docs/rem_report.pdf

EPA. 2004a. *Cleaning Up the Nation's Waste Sites: Markets and Technology Trends.*
http://www.clu-in.org/download/market/2004market.pdf

EPA. 2004b. *Strategy to Ensure Institutional Control Implementation at Superfund Sites.* http://www.epa.gov/superfund/ action/ic/icstrategy.pdf

EPA. 2004c. *Superfund: Building on the Past, Looking to the Future.* http://www.epa.gov/superfund/action/120day/pdfs/study/120daystudy.pdf

EPA. 2004d. *Updating Remedy Decisions at Select Superfund Sites: Summary Report FY 2002 and FY 2003.* http://www.epa.gov/superfund/programs/reforms/docs/urd02-03.pdf

EPA. 2005a. *About Brownfields.* http://www.epa.gov/brownfields/about.htm (accessed May 4, 2005).

EPA. 2005b. *About CLU-IN.* http://www.clu-in.org/about (accessed April 27, 2005).

EPA. 2005c. *Background on the Libby Asbestos Site.*
http://www.epa.gov/region8/superfund/libby/background.html (accessed April 2, 2005).

EPA. 2005d. *Data Sharing Pilots.* http://www.epa.gov/superfund/action/ic/datashar/index.htm (accessed April 27, 2005).

EPA. 2005e. *NPL Site Totals by Status and Milestone.*
    http://www.epa.gov/superfund/sites/npl/index.htm (accessed October 3, 2005).
EPA. 2005f. *Reforms by Round.* http://www.epa.gov/superfund/programs/reforms/byround.htm
    (accessed April 27, 2005).
EPA. 2005g. *Return to Use: An Initiative to Remove Barriers to Reuse at Superfund Sites.*
    http://www.epa.gov/superfund/programs/recycle/rtu/index.htm (accessed October 6,
    2005)
EPA. 2005h. *Return to Use Demonstration Projects.*
    http://www.epa.gov/superfund/programs/recycle/rtu/demos.htm (accessed April 5,
    2005).
EPA. 2005i. *Superfund Information Systems: Cerclis Database.*
    http://cfpub.epa.gov/supercpad/cursites/srchsites.cfm (accessed October 9, 2005).
EPA. 2005j. *Superfund Innovative Technology Evaluation.* http://www.epa.gov/ORD/SITE
    (accessed April 27, 2005).
EPA. 2005k. *Superfund Redevelopment Program: At a Glance.*
    http://www.epa.gov/superfund/programs/recycle/index2.htm (accessed October 3,
    2005).
EPA. 2005l. EPA, Region 7. *Times Beach Site, Times Beach, Missouri.*
    http://www.epa.gov/region7/cleanup/npl_files/mod980685226.pdf (accessed April 2,
    2005).
Esty, Daniel C. 1996. "Revitalizing Environmental Federalism." *Michigan Law Review* 95:
    570-653.
Federal Register. 1990. *National Oil and Hazardous Substances Pollution Contingency Plan.*
    55 Fed. Reg. 8,669 (March 8, 1990) (to be codified at 40 C.F.R. pt. 300).
Ferguson, Margaret Calder. 2002. "Evaluation of Remediation Technologies for Various
    Contaminants Found on Superfund Sites." April 2002. Unpublished manuscript on
    file with author.
Freeman, Jody. 1997. "Collaborative Governance in the Administrative State." *U.C.L.A. Law
    Review* 45: 1-98.
Front Royal-Warren County Economic Development Authority. 2005. *Avtex Redevelopment.*
    http://www.wceda.com/newpage12.htm (accessed October 4, 2005).
Gayer, Ted, James T. Hamilton, and W. Kip Viscusi. 2000. "Private Values of Risk Tradeoffs
    at Superfund Sites: Housing Evidence on Learning about Risk." *Review of Economics
    and Statistics* 82: 439-451.
"Green Light for Superfund Site." 2004. *Brownfield News.* December 2004.
    http://www.brownfieldnews.com/archive/december/V8I5_western_utah.htm
Gunderson, Lance H., C. S. Holling, and Stephen S. Light. 1995. "Barriers Broken and Bridges
    Built: A Synthesis." In *Barriers and Bridges to the Renewal of Ecosystems and
    Institutions,* Lance H. Gunderson, C. S. Holling, and Stephen S. Light, eds., 489-532.
    New York: Columbia University Press.
Hagerstown Land Use Committee, E² Inc., and University of Virginia's Institute for
    Environmental Negotiation. 2003. *Central Chemical Superfund Redevelopment
    Initiative Pilot Project: Project Report.*
    www.virginia.edu/ien/HagerstownLUCFinalReport.pdf
Hagerstown Minutes. 2004. Mayor & City Council, Hagerstown, Maryland, 60th Sess., Regular
    Session Minutes (May 25, 2004).
    http://www.hagerstownmd.org/CityGov/councilminutes.asp
Hennelly, Robert. 2003. "Superfund Heading for a Super Crisis." *New Jersey Representative.*
    January–February.
Hersh, Robert, Katherine Probst, Kris Wernstedt, and Jan Mazurek. 1997. *Linking Land Use
    and Superfund Clean Ups: Uncharted Territory.* Washington, DC: Resources for the
    Future. www.rff.org/Documents/RFF-RPT-landuse.pdf

Holling, C. S. 1978. *Adaptive Environmental Assessment and Management.* Chichester, New York: Wiley.

Holling, C. S. 1992. "Cross-Scale Morphology, Geometry, and Dynamics of Ecosystems." *Ecological Monographs* 62: 447-502.

Jiggins, Janice, and Niels Röling. 2002. "Adaptive Management: Potential and Limitations for Ecological Governance of Forests in a Context of Normative Pluriformity." In *Adaptive Management: From Theory to Practice,* James Oglethorpe, ed., 94-103. Cambridge: World Conservation Union.

John, DeWitt. 1999. "Good Cops, Bad Cops." *Boston Review,* October–November: 19.

Karkkainen, Bradley C. 2003. "Adaptive Ecosystem Management and Regulatory Penalty Defaults: Toward Bounded Pragmatism." *Minnesota Law Review* 87: 943-998.

Laws, Elliot P. 1995. Memorandum from Elliott P. Laws, Assistant Administrator, U.S. EPA, to Regional Directors. *Land Use in the CERCLA Remedy Selection Process* (May 25, 1995). http://www.epa.gov/swerosps/bf/pdf/land_use.pdf

Leahy, Leslie. 2004. Personal communications with Leslie Leahy, Superfund Community Involvement and Outreach Branch, U.S. EPA, Washington, DC, regarding EPA Community Advisory Group List as of 10/05/04 (October 27, 2004). On file with author.

Lee, Kai N. 1993. *Compass and Gyroscope: Integrating Science and Politics for the Environment.* Washington, DC: Island Press.

Lee, Kai N. 1999. "Appraising Adaptive Management." *Conservation Ecology* 3. http://www.ecologyandsociety.org/vol3/iss2/art3/index.html

Leopold, Aldo. 1949. *A Sand County Almanac* and *Sketches Here and There.* New York: Oxford University Press.

Lindblom, Charles E. 1959. "The Science of 'Muddling Through.'" *Public Administration Review* 19: 79-88.

Linkov, Igor, A. Verghese, S. Jamil, T.P. Seager, G. Kiker, and T. Bridges. 2004. "Multi-Criteria Decision Analysis: A Framework for Structuring Remedial Decisions at Contaminated Sites." In *Comparative Risk Assessment and Environmental Decision Making,* Igor Linkov and Abou Bakr Ramadan, eds., 15-34. Boston: Kluwer. www.environmentalfutures.org/Images/ArmyPaper_Oct31.pdf

Lord, Charles P., Eric Strauss, and Aaron Toffler. 2003. "Natural Cities: Urban Ecology and the Restoration of Urban Ecosystems." *Virginia Environmental Law Journal* 21: 317-385.

Luftig, Stephen D. 1999. Memorandum from Director, Office of Emergency and Remedial Response, U.S. EPA, to Superfund National Policy Managers Regions 1–10, Issuance of Final Guidance: *Ecological Risk Assessment and Risk Management Principles for Superfund Sites* (October 7, 1999). http://www.epa.gov/oswer/riskassessment/pdf/final10-7.pdf

Luftig, Stephen D., and Barry N. Breen. 1996. Memorandum from Stephen D. Luftig, Director, Office of Emergency and Remedial Response, and Barry N. Breen, Director, Office of Site Remediation Enforcement. *Superfund Reforms: Updating Remedy Decisions* (September 27, 1996). http://www.epa.gov/superfund/resources/gwdocs/updating.pdf

McClelland, Gary H., William D. Schulze, and Brian Hurd. 1990. "The Effect of Risk Beliefs on Property Values: A Case Study of a Hazardous Waste Site." *Risk Analysis* 10: 485-497.

McCluskey, Jill J. and Gordon C. Rausser. 2003. "Stigmatized Asset Value: Is It Temporary or Long-Term?" *Review of Economics & Statistics* 85: 276-285.

McDaniels, Timothy L., and Robin Gregory. 2004. "Learning as an Objective within a Structured Risk Management Decision Process." *Environmental Science and Technology* 38: 1921-1926.

Meffe, Gary K., Nielsen, Larry A., Knight, Richard L., and Schenborn, Dennis A., eds. 2002. *Ecosystem Management: Adaptive, Community-Based Conservation*. Washington, DC: Island Press.

Moir, W. H., and W. M. Block. 2001. "Adaptive Management on Public Lands in the U.S.: Commitment or Rhetoric?" *Environmental Management* 28: 141-148.

Mundy, Bill. 1992. "Stigma and Value." *Appraisal Journal* 60: 7-13.

NACEPT. *See* Superfund Subcommittee, National Advisory Council for Environmental Policy and Technology.

National Research Council. 1992. *Restoration of Aquatic Ecosystems: Science, Technology and Public Policy*. Washington, DC: National Academy Press.

National Research Council. 2003. *Environmental Cleanup at Navy Facilities: Adaptive Site Management*. Washington, DC: National Academy Press.

National Research Council. 2003. *Long-Term Stewardship of DOE Legacy Waste Sites: A Status Report*. Washington, DC: National Academy Press.

Norton, Bryan G., and Anne Steinemann. 2001. "Environmental Values and Adaptive Management." *Environmental Values* 10: 473-506.

NRC. *See* National Research Council.

Oates, Wallace E. 1972. *Fiscal Federalism*. New York: Harcourt Brace.

Oates, Wallace E. 1997. "On Environmental Federalism." *Virginia Law Review* 83: 1321-1329.

Olson, Mancur. 1965. *The Logic of Collective Action: Public Goods and the Theory of Groups*. Cambridge, MA: Harvard University Press.

Prichard, Jr., Lowell, Carl Folke, and Lance Gunderson. 2000. "Valuation of Ecosystem Services in an Institutional Context." *Ecosystems* 3: 36-40.

Probst, Katherine N., D. Konisky, R. Hersh, M. Batz, and K. Walker. 2001. *Superfund's Future: What Will It Cost?* Washington, DC: Resources for the Future.

Probst, Katherine N., and Diane Sherman. 2004. *Success for Superfund: A New Approach for Keeping Score*. Washington, DC: Resources for the Future. http://www.rff.org/documents/RFF-RPT-SuperfundSuccess.pdf

Quarles. 2003. Telephone interview with John Quarles, Senior Counsel, Morgan Lewis. Washington, DC, November 10, 2003.

Reed, Larry. 2001. Memorandum from Larry Reed, Acting Director, Office of Emergency and Remedial Response, U.S. EPA, to Superfund National Policy Managers Regions 1–10. *Reuse Assessment: A Tool to Implement the Superfund Land Use Directive* (June 4, 2001). http://www.epa.gov/superfund/resources/reusefinal.pdf

Revesz, Richard L. 1992. "Rehabilitating Interstate Competition: Rethinking the 'Race-to-the-Bottom' Rationale for Federal Environmental Regulation." *New York University Law Review* 67: 1210-54.

Sabatier, Paul A. 1988. "An Advocacy Coalition Framework of Policy Change and the Role of Policy-Oriented Learning Therein." *Policy Science* 21: 129-168.

Sarnoff, Joshua D. 1997. "The Continuing Imperative (But Only From a National Perspective) for Federal Environmental Protection." *Duke Environmental Law and Policy Forum* 7: 225-319.

Strasser, Kurt A., and William Breetz. 2003. "Benefits of a Uniform State Law for Institutional Controls." In *Implementing Institutional Controls at Brownfields and Other Contaminated Sites*, Amy L. Edwards, ed., 31-37. Chicago, IL: American Bar Association.

Sunstein, Cass R. 2002. *The Cost-Benefit State: The Future of Regulatory Protection*. Chicago, IL: American Bar Association.

Superfund Subcommittee, National Advisory Council for Environmental Policy and Technology (NACEPT). 2004. *Final Report* (April 2004). http://www.epa.gov/oswer/docs/naceptdocs/NACEPTsuperfund-Final-Report.pdf

Suter II, Glenn W., Rebecca A. Efroymson, Bradley E. Sample, and Daniel S. Jones. 2000. *Ecological Risk Assessment for Contaminated Sites.* Boca Raton, FL: Lewis Publisers.

Torrell, Elin. 2000. "Adaptation and Learning in Coastal Management: The Experience of Five East African Initiatives." *Coastal Management* 28: 353-363.

UECA. *See* (proposed) Uniform Environmental Covenants Act.

Uniform Environmental Covenants Act (proposed). 2003. http://www.law.upenn.edu/bll/ulc/ueca/2003final.pdf

Walker, Katherine D., March Sadowitz, and John D. Graham. 1995. "Confronting Superfund Mythology: The Case of Risk Assessment and Risk Management." In *Analyzing Superfund: Economics, Science, and Law,* Richard L. Revesz and Richard B. Stewart, eds., 25-53. Washington, DC: Resources for the Future.

Walters, Carl. 1986. *Adaptive Management of Renewable Resources.* New York: MacMillan.

3

ADAPTIVE MANAGEMENT:
A Review and Framework for Integration with Multi-Criteria Decision Analysis

F. Kyle Satterstrom and Igor Linkov[1]
Cambridge Environmental, Inc.

Gregory Kiker and Todd Bridges
U.S. Army Engineer Research and Development Center

Marc Greenberg
Office of Solid Waste and Emergency Response, U.S. Environmental Protection Agency

ABSTRACT

Conventional management practice has traditionally focused on finding the best available policy option. Given the uncertain and ever-changing nature of environmental and social conditions, however, this has proven to be a daunting task. Adaptive management acknowledges that no single policy can be selected, but rather a set of alternatives should be dynamically tracked to reveal the best course of action at any given time. Although adaptive management concepts were introduced more than twenty years ago, their implementation is generally piecemeal; fully developed adaptive planning and procedural frameworks have been limited to large-scale projects in long-term natural management, where uncertainty is often overwhelming. Nevertheless, even conventional managers of smaller projects are confronted with the same problems and may go through the frustrating experience of changing their management strategy when it fails. In this chapter, we review regulatory policies and adaptive management implementations across a wide range of projects and application areas. Our review indicates a need to integrate adaptive management with a set of decision-making tools that will allow it to build on current management approaches. We are thus proposing a solution in which we choose a strong adaptive management framework, for which there exist many support tools in the literature, and integrate it with multi-criteria

---

[1]   Corresponding author, linkov@CambridgeEnvironmental.com

decision analysis (MCDA) as a method for dealing with uncertainty when selecting a management option. The two methods complement each other and fit together smoothly, forming a comprehensive management framework.

INTRODUCTION

Choosing the best possible environmental management strategy is extremely important, yet it can be a complex and difficult problem. The environmental system may contain multitudes of species and a variety of landscapes while simultaneously straining under the pressure of human development. On top of this, the system can be dynamic and our knowledge of it highly uncertain – but in traditional management, goals are set and held static. Although different strategies are considered as possible ways to attain existing goals, it is assumed that sufficient knowledge exists for picking one "best" option. Once a strategy is selected as the optimal approach and implemented, its performance may or may not be monitored closely. At some point, the strategy will likely be evaluated, and if the action is perceived to have failed, a different strategy may be substituted, or further studies may be conducted. Either management response will add unanticipated expenses. The goals themselves are not often reconsidered, and in such a framework any change in the management strategy or admission of uncertainty about the system being managed is prone to be interpreted as weakness or failure. The result can be a management strategy which is ill-fitted to the site, even if it is considered optimal at the outset.

The management decision paradigm currently implemented at Superfund sites (EPA 2005) is an example of an area which could benefit from the application of adaptive management. The Superfund process can be predominantly front- and end-loaded (see Figure 3.1a). After initial scoping, remedial investigation (RI) and a feasibility study (FS) are conducted to examine different managerial strategies. Risk analysis and modeling conducted within the RI/FS process partly influence the choice of which strategy to implement, but other factors (political, social, economic, or technical) may also greatly affect the decision process. In this ad hoc environment, the decision-maker may be forced to draw upon various heuristics (i.e., rules of thumb) to reduce the complexity of the decision. Thus, even when managers spend large amounts of time attempting to determine an optimal management strategy, they are still constrained to a framework which may not fully consider all available information. Further, once a decision is made and the management plan is implemented, it is not easily modified (for example, if new data that would significantly influence the decision are obtained) until the site's five-year review. Even at this point the initial decision process may have been difficult enough to discourage a comprehensive review.

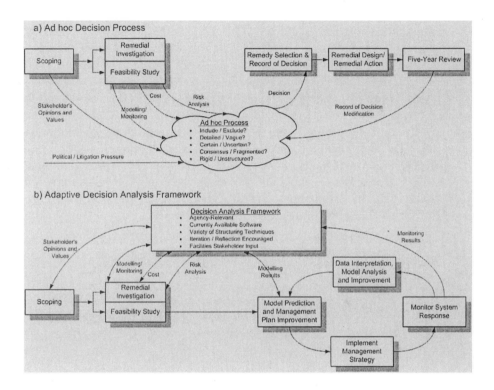

Figure 3.1. Ad hoc decision process and adaptive decision analysis framework.

Current developments in the field of environmental management have resulted in many attempts to improve the environmental management process. The Interstate Technology and Regulatory Council, for example, has developed strategies for optimizing the Superfund remediation process (ITRC 2004) that provide guidelines for reevaluation and adjustment of selected remedial alternatives. Similar guidelines have recently been developed by other agencies and groups, such as a comprehensive monitoring framework (EPA 2004a). However, even though these documents guide one through the process of incorporating new information into the decision process, they still presume that the best available alternative will be selected based on the changes that occurred after a previous search for an optimal approach.

Adaptive management, by contrast, acknowledges at the outset that uncertainty is inherent in any natural system. It seeks to minimize this uncertainty by learning about the system being managed. The basic process is straightforward: when managing any system, one chooses an action, monitors the effects of the action, and adjusts the action based on the monitoring results. There are two types of adaptive management: passive (Figure 3.2b) and active (Figure 3.2c) (Wilhere 2002). Both processes begin by setting goals, modeling the system, and selecting and implementing a management strategy. Passive adaptive management involves implementing one management strategy at a time, whereas in active adaptive management multiple potential alternatives are examined alongside a control to isolate factors which affect the system. The managed ecosystem is then monitored to collect decision-oriented data regarding the effects of the management strategy on the system. The results of the monitoring affect model estimation and parameter values, and the management strategy is evaluated and adjusted as a result. During the adaptive management process, in contrast to traditional management, change is expected and welcomed, learning is emphasized, and there is flexibility in that the goals and objectives may occasionally need to be modified based on the performance of the management strategy.

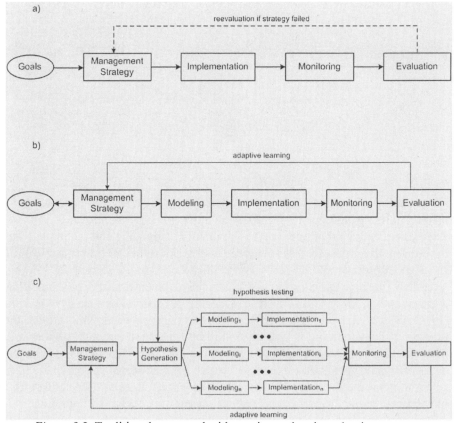

Figure 3.2. Traditional compared with passive and active adaptive management.

Six elements of adaptive management set out by the National Research Council (2004) provide a comprehensive framework for the adaptive management process:

- First, the process must have *management objectives which are regularly revisited and accordingly revised.* Stakeholders must agree on what the basic objectives are, and the project's objectives should be reviewed when new information becomes available.
- Second, *a model of the system(s) being managed* is necessary to provide an understanding of the system, both for management option generation and also for result explanation. It, too, should be updated when new information is obtained.
- Third, stakeholders should generate *a range of management choices.* Implementing multiple numbers of these simultaneously, along with a control, can provide information similar to that of a scientific experiment.
- Fourth, the process requires *monitoring and evaluation of outcomes.* Monitoring enables the manager to distinguish which strategies worked better than others.
- Fifth, having obtained this information, an adaptive management process should have *a mechanism(s) for incorporating learning into future decisions.* Learning about the system in order to better manage it is, after all, the point of the whole process.
- Finally, management should be undertaken with *a collaborative structure for stakeholder participation and learning.* All affected parties need to feel represented and gain information throughout the process.

These elements form a thorough set of criteria which, if followed, provide an effective framework for adaptive management.

If the Superfund program were conducted under an adaptive management paradigm, the acceptance of uncertainty at the stage of the RI/FS would result in a set of management scenarios that would be implemented and adapted based on changing environmental and evolving social trends (see Figure 3.1b). Such an approach allows the manager to learn about the system being managed and to modify the current strategy based on new knowledge. Our review will show, however, that the field of environmental management is far from accepting and using adaptive management approaches. Although adaptive management is recognized and even recommended by many state and government agencies (as we show below), adaptive management applications vary widely in their implementation of the concept and there is no framework that robustly incorporates the whole of adaptive management in environmental practice.

We propose to combine adaptive management philosophy with decision analysis tools (Linkov et al. 2004; Linkov et al. 2006a; Linkov et al. 2006b; Kiker et al. 2005) to offer a structured framework for selecting the best managerial strategy. This framework is built on current optimization models, but it applies them in an iterative manner wherein management plans are reexamined on a continual basis rather than only when they fail. Adaptive management is treated as a planning and uncertainty-mitigation tool as well as a procedural tool in this framework. The continual cycle of prediction, implementation, monitoring, and evaluation allows new information to be incorporated into the decision-making process (Figure 3.1b). This cycle, along with the decision analysis framework, gives structure to the process and ensures that all available information is considered.

ADAPTIVE MANAGEMENT IN GOVERNMENT

Regulatory agencies in the United States and around the world recommend adaptive management, but agencies often implement or emphasize different parts of the adaptive management process. Table 3.1 (Appendix) summarizes the role of adaptive management in a number of federal and state agencies in the US and other countries; a detailed review is provided below.

US Environmental Protection Agency

Adaptive management has been recommended for use in many of the EPA's regulatory programs. Regulating the total maximum daily loads (TMDLs) of pollutants in various media by the Office of Water has probably resulted in the largest number of applications. The EPA has also developed the Triad Approach to support adaptive field decisions in order to manage uncertainty. The triad streamlines the sampling, analysis, and data management activities conducted during site assessment, characterization, and cleanup (EPA 2003; Crumbling et al. 2004). The EPA defines adaptive management as an "[a]pproach where source controls are initiated while additional monitoring data are collected to provide a basis for future review and revision of the TMDL (as well as management activities)" (EPA 2004b). Adaptive management is discussed briefly in both the EPA's *Protocol for Developing Sediment TMDLs* (EPA 1999c) as well as its *Protocol for Developing Pathogen TMDLs* (EPA 2001a). Both protocols emphasize careful monitoring of pollutant levels, but neither develops a complete adaptive management framework.

Two additional EPA programs advocate passive adaptive management. The Office of Wetlands, Oceans, and Watersheds' Watershed Analysis and

Management (WAM) Guide for Tribes utilizes a passive adaptive management approach emphasizing planning, monitoring, evaluating, and adjusting, but it does not consider a range of options or incorporating stakeholder input (EPA 2000). The Office of Solid Waste and Emergency Response, meanwhile, recommends adaptive management and its basic elements (EPA 1999b) as a model for sustainable brownfields (i.e., land whose development may be hindered by the presence of hazardous substances) and as a framework for monitoring plan development and implementation at hazardous waste sites (EPA 2004a). Even though the guidance lists important features of adaptive management, it does not provide application examples or other means to help users implement it.

The EPA has implemented adaptive management concepts in a few notable projects. Its Mississippi River Basin project is designed to use adaptive management to assess the causes and consequences of hypoxia in the Gulf of Mexico (EPA 1999a; EPA 2001b). The project utilizes models and monitoring to reduce the uncertainty surrounding the biochemical mechanisms of hypoxia. However, the project does not generate or employ a range of options, nor does it revisit objectives or emphasize stakeholder collaboration. Another application of adaptive management is the joint EPA/Environment Canada Lake Superior Lakewide Management Plan (EPA 2002a). The Binational Executive Committee recommends the use of adaptive management throughout the project – it emphasizes an iterative process which gains knowledge about the ecosystem, accepts public input, and revisits and revises the objectives of the project based on the entire body of knowledge current at the time. However, this interpretation of adaptive management is a periodic refining of management plans based on new information and public input – a much less structured approach than one involving modeling, active option generation, and comprehensive monitoring. These examples underscore the range of variability in adaptive management projects – whereas the Mississippi River Basin project emphasizes the modeling and monitoring aspects of adaptive management but lacks stakeholder involvement, the Lake Superior project solicits public input but does not emphasize modeling or monitoring.

Department of Energy

Adaptive management has also received attention within the Department of Energy (DOE). It was given a cursory mention in the DOE's 1996 Comprehensive Land-Use Planning Process Guide (DOE 1996), where it is mentioned as an iterative decision-making framework which can take into account new information. The adaptive management section of this guide is very brief, and the DOE's NEPA Task Force report, "Modernizing NEPA Implementation" (DOE 2003), finds that the department could employ adaptive management to a much greater extent. The adaptive management program espoused by the Task Force is referred to as the "predict-mitigate-implement-monitor-adapt" model. As with many other government documents, heavy emphasis is placed on monitoring, and stakeholder collaboration is discussed as if the idea were new.

One area where the DOE is seeking to put adaptive management into practice is managing the environmental effects of hydroelectricity production. For example, the adaptive management plan for hydroelectricity production and salmon conservation in the Columbia River basin of Washington state (DOE 2002) emphasizes the use and continual refinement of advanced mathematical models to interpret and predict environmental change, as well as to evaluate trade-offs and possible management options. On a more general level, the Federal Energy Regulatory Commission (FERC) under the DOE has adopted adaptive management as a way to deal with uncertainty about the possible environmental effects of a power plant's construction. If a potential power installation applies for a new hydropower license, FERC may choose to expedite license issuance under the condition that adaptive management be used at the site (FERC 2000). The ecological effects of power production installations are not always known; thus, the DOE has moved to manage the related mitigation measures adaptively.

Department of the Interior

Adaptive management is being integrated into procedures of the Department of the Interior (DOI). The DOI's Departmental Manual (DOI 2004) encourages the department's bureaus to build adaptive management practice into their proposed actions and NEPA compliance activities, including training personnel in the concept. In addition, heads of Bureaus are required to use adaptive management to comply with NEPA requirements.

Adaptive management has been implemented in many research projects and site-specific applications by DOI scientists. The Southeastern Adaptive Management Group (SEAMG 2004), for example, was created in 2001 in order to help the agency achieve a better science-based approach to wildlife conservation and management. In addition, the Bureau of Land Management has advocated adaptive management in monitoring its use authorizations for oil and gas (BLM 2003), and the National Parks Service has developed a table containing policy options for various issues in the adaptive management of Yellowstone and Grand Teton national parks (NPS 2002).

The Bureau of Reclamation has also implemented adaptive management, in its Battle Creek, California, project to restore salmon populations (USBR 2001). Their adaptive management diagram includes, albeit not always in the same terms, each of the six elements of adaptive management that we have chosen to emphasize in this paper. The document emphasizes monitoring, even detailing exactly which characteristics of the fish need to be recorded. Besides the Bureau of Reclamation, other entities including the Pacific Gas and Electric Company, National Marine Fisheries Service, the U.S. Fish and Wildlife Service, and the California Department of Fish and Game are involved in the administration of the project, and they apply adaptive management in their attempt to restore and enhance habitat while minimizing the loss of hydroelectric energy production by dams on the river.

To cite a TMDL example in the DOI, the U.S. Geological Survey has promoted the use of an adaptive management approach to help wastewater treatment plants use offsets to meet mercury regulatory standards in California's Sacramento River Watershed (USGS 2004). The plan includes stakeholder involvement, comprehensive system modeling, and methodology optimization, but little of the iterative nature of adaptive management. Monitoring is implicit but not well described. Similarly, the Water Resources Research Institute at North Carolina State University called for adaptive management of the local Neuse River Estuary, again placing an emphasis on modeling while neglecting other aspects of adaptive management.

NOAA

The National Oceanic and Atmospheric Administration (NOAA) utilizes adaptive management, especially in their coastal management and coastal habitat restoration activities (NOAA 2004c). The process espoused is again a passive one, involving an iterative, five-step cycle: plan, act, monitor, evaluate, and adjust (NOAA 2004a). NOAA emphasizes the monitoring and evaluation steps of adaptive management – Sutinen (2000) developed a monitoring and assessment framework for the agency – but the degree to which their projects include other aspects, such as modeling or revisitation of objectives, can vary widely. NOAA has used adaptive management in projects such as shore restoration in the Pacific Northwest, the restoration of native plant species in a Rhode Island marsh, and in larger projects such as wetlands protection in Louisiana, where there is again an emphasis on learning through monitoring (NOAA 2004b).

The National Marine Fisheries Service (NMFS) created the Plan for Analyzing and Testing Hypotheses (PATH) in 1995 as a way to reconcile the variation among different entities' analyses of declining salmon stocks in the Snake River basin (Marmorek and Peters 2001). PATH focused on interagency collaboration, including scientists from the University of Washington, the Oregon Department of Fish and Wildlife, and the Bonneville Power Administration. The agencies developed common data sets, and worked toward resolving such issues as differing objectives among participants, lack of trust, and lack of understanding of the underlying assumptions of others' models. The program was successful in that it allowed for models to be more consistent with observed data and improved prediction of the effects of future management actions on salmon stocks. PATH implemented many adaptive management ideas, such as stakeholder collaboration (although limited to scientists), utilization of models and a focus on the reduction of uncertainty, and an iterative structure. It concluded that dam removal was the surest way to restore salmon runs. Unfortunately, funding for PATH was not renewed for the year 2000.

USDA

The Department of Agriculture also recognizes that ecosystem knowledge is incomplete. A USDA Natural Resources Conservation Service (NRCS) handbook recommends adaptive management to deal with uncertainty in conservation planning, defining it as "the process of using monitoring, evaluation and experimentation to provide information that can be used in future management decisions" (USDA 2003a). Another handbook section discusses monitoring, emphasized because it is seen as the step which provides the information gained during adaptive management (USDA 2003b). An NRCS work plan for adaptive management in the Klamath River basin reflects this view – policy application, monitoring, and policy adjustment are the only specific steps given for adaptive management.

Department of Defense

The Department of Defense (DOD) is also adopting adaptive management concepts. A National Research Council book (NRC 2003) provides a thorough overview of how to apply adaptive management to the remediation of contaminated sites, tailored to the cleanup of Navy facilities. Their framework emphasizes experimentation, monitoring and evaluation, and public involvement. Adaptive site management, carried out according to this approach, shares many of the elements that we have adopted as key to adaptive management. The U.S. Army Corps of Engineers (USACE) has also begun utilizing adaptive management principles in its water resources projects, such as the Florida Everglades, Missouri River Dam and Reservoir System, Upper Mississippi River, and coastal Louisiana (NRC 2004), although specific policy and framework development is still being formulated.

Canada, Europe, and Private Organizations

Varying aspects of adaptive management have also been implemented in Canada. Fisheries and Oceans Canada has advanced a socioeconomic framework for ecosystem-based fisheries management (Rudd 2004), and two of the recommendations from Health Canada's report on the Health Impacts of the Greenhouse Gases Mitigation Measures include adaptive management. The panel advises that emissions trading could result in health problems in urbanized areas, and it recommends that adaptive management be used to learn about the problem and find a flexible solution to it. The panel is not clear about what it considers the specific aspects of adaptive management to be, but it does recommend that stakeholders be involved in the process of learning about the system (CCHO 2000).

Environment Canada has also employed elements of adaptive management. Their Biodiversity Convention Office, which emphasizes learning through monitoring, uses adaptive management to compensate for resource managers' lack of complete knowledge and understanding of complex ecosystems and ecosystem structure and functions (Environment Canada 2004a). The Environmental Assessment Best Practice Guide for Wildlife at Risk in Canada recommends that adaptive management be applied to managing and learning about wildlife (Environment Canada 2004b), and Canada's National Wildlife Disease Strategy contains recommendations such as the development of adaptive management strategies that are based on principles of ecology and conservation (Environment Canada 2004c), but neither project develops their general adaptive management approach into a structured framework.

On a provincial level, the British Columbia Forest Service is currently implementing many pilot adaptive management projects, and they do have guides to adaptive management published locally (Nyberg 1999; Taylor 1997). Their projects include adaptive management of livestock grazing, riverbanks and streambanks, forest recreation sites, and grizzly bear habitat, among others (BCFS 2000). The grizzly bear project is intended to develop forest systems that can provide forageable land for grizzlies while at the same time producing a sustainable timber output. To achieve this main objective, forest managers have produced trial tree clusters in which they vary the gap size between trees, the number of trees per cluster, and the forage plant species in the gaps between trees. They also have secondary objectives, including evaluating the effect of varying gap size on light levels in the gaps. The BC Forest Service is a strong example of the integration of adaptive management into agency operations.

Also in British Columbia, the province's Ministry of Sustainable Resource Management has recommended implementation of adaptive management in its North Coast Land and Resource Management Plan (BCMSRM 2004). Although the plan does not contain an explicit adaptive management framework, it does differentiate between passive and active adaptive management approaches. The plan states that both are relevant to the North Coast plan, and it advises developers to implement an adaptive management process. The plan includes a method for making amendments to previously implemented policies when new information becomes available.

In the European Union, by contrast, there are a number of areas where adaptive management *could* be applied. The Commission of the European Communities, for example, recognizes many problems with the current management of fisheries stocks (CEC 2001). Many of the stocks are over-harvested, there is insufficient monitoring of the status of the fisheries, and knowledge about the workings of marine ecosystems as well as the effects of fishing on them is limited. The World Wildlife Federation (WWF) has addressed this problem specifically, and they recommend the use of what are effectively adaptive management principles in the governance of European fisheries. The WWF calls for ecosystem-based approaches, smaller fishing fleets, and increased stakeholder involvement, among other practices (WWF 2001a).

The Biodiversity Support Program, with roots in The Nature Conservancy as well as the World Wildlife Federation, has published a general guide to adaptive management which defines it as "the integration of design, management, and monitoring to systematically test assumptions in order to adapt and learn" (WWF 2001b). The guide explicitly relates adaptive management to the scientific method and emphasizes learning and reducing uncertainty. Between the adaptive management steps given, as well as the principles given for carrying them out, the guide addresses all of the adaptive management elements that we have adopted.

The Global Environment Facility (GEF), an intergovernmental environmental group, has supported many environmental projects across dozens of nations since its founding in 1991. The GEF endorses continuous learning, and it has an entire unit, the GEF Monitoring and Evaluation Unit (GEFME), which assesses GEF programs and projects and reports the results and lessons, providing a feedback mechanism for learning from past projects (GEF 2004). Much as with government agencies the world over, however, individual projects do not necessarily follow an adaptive management framework.

The RAND Corporation, a global non-profit research institution, espouses the idea of adaptive management and adaptive decision strategies as a way to help conduct long-term policy analysis (Lempert et al. 2003). Like the proposal for the Columbia River basin, the RAND discussion centers on how computers and computer modeling can help decision-makers plan amidst deep uncertainty. When they mention adaptive management, they do so in a general fashion, comparing adaptive learning to free economic markets and representative governments; they do not go into any specifics of the adaptive management process.

Throughout this review, it is evident that government and other agencies do not necessarily have a systematic approach to adaptive management. One study may emphasize modeling, another monitoring, and so forth. Many current government documents emphasize monitoring and adjusting, and although they may even mention concerns such as stakeholder involvement, they do not always view these as a necessary component of adaptive management. Clearly, managers across many different agencies could benefit from a clear adaptive management framework and related tools.

ADAPTIVE MANAGEMENT TOOLS AVAILABLE IN THE LITERATURE

To explore the tools available for supporting each of the six elements of the adaptive management process identified by NRC (2004), we conducted a literature review which reveals the existence of hundreds of peer-reviewed papers that deal with adaptive management and its application in different areas of environmental research and administration. Most of these papers address just a few elements of adaptive management and thus do not provide a coherent application framework. We provide tables summarizing the literature by subject area (see Table 3.2 in the Appendix for application papers and Table 3.3 for review papers), but we discuss the literature according to each of the six elements below. Throughout the review, we encountered papers that dealt with more than one of the six elements. We discuss these below under their primary concern, but their other identified elements are reflected in the tables.

1. Management Objectives Which are Regularly Revisited and Accordingly Revised

The first element of adaptive management is a regular revisitation of a project's objectives. Interestingly, not many adaptive management papers explicitly discuss the updating and revision of objectives upon the acquisition of new information. Many, in fact, take it for granted that their objectives are static goals. These papers have actually missed an important aspect of the adaptive management process as laid out by the NRC (2004). Goals and objectives need to be based on all the information available to stakeholders, and as this pool of information grows, objectives should be revised. The National Research Council (NRC 2003) also incorporates this step in their framework for the remediation of contaminated Navy sites. Their book includes sections on monitoring and data analysis to support experimentation, evaluation, and public participation – but its revisiting of objectives is an important aspect of its plan which should not be overlooked.

The objectives themselves can be formulated in different ways. Goals may be the restoration of an ecosystem or the preservation of a species, but they are not always phrased in this language. For example, techniques of risk assessment naturally couple with adaptive management, where reduction of risk is a main endpoint by which the success of a management plan is measured. Gentile et al. (2001) discuss the integration of ecological risk assessment and adaptive management in the context of Florida Everglades restoration.

2. A Model of the System(s) Being Managed

Modeling tools are integral to many adaptive management processes. They provide a basis for understanding why change occurs in the environment being managed, and they can also be used to predict the effects of possible strategies during policy selection. For example, Bearlin et al. (2002) use a stochastic population model for trout cod while choosing an adaptive management strategy for their reintroduction, and van Damme et al. (2003) implement forest management models consisting of four modules – timber supply, biodiversity, fire, and water – when generating results to adjust the management scenarios and choose one for implementation. Van Staden et al. (2004) model the distribution of two fungal pathogens as they consider the possible effects of climate change, pests, and diseases on the adaptive management of commercial forests in South Africa.

Modeling has also been used to examine the differences between passive and active adaptive management. Johnson and Williams (1999) discuss the benefits of adaptive waterfowl management in reducing the uncertainty in waterfowl population dynamics, and they draw upon passive and active adaptation algorithms in their discussion of waterfowl harvest regulations. Bundy (2004) also compares passive and active management strategies as he examines four different models of the San Miguel Bay fishery in the Philippines. Based on the modeling, he concludes that although there would be little to gain from pursuing active adaptive management over passive adaptive management in this particular case, an adaptive management strategy would certainly be preferable over a static strategy with little monitoring or learning.

Just as they can be employed for many reasons, models can be built in a variety of ways. Bliss et al. (1997) utilize a method called the Delphi technique to elicit knowledge from experts on Southern bottomland hardwood ecosystems. This technique is an iterative survey process in which experts are surveyed for information and their questionnaire answers are used to design subsequent questionnaires. The survey results are used to design quantitative and qualitative aspects of a computer model of a bottomland hardwood ecosystem. Similarly, Bunch and Dudycha (2004), who use adaptive management principles in planning the cleanup of the Cooum River in India, build their models of various management scenarios using input gained through stakeholder workshops.

While mathematical models are preferred when possible, conceptual models are also desirable because adaptive managers often deal with substantially uncertain situations. Thom (1997; 2000) employs the concept of a system-development matrix in the management of coastal ecosystems. The matrix is used to describe the current state of the ecosystem and to suggest actions to adaptively manage the system toward its desired state. Gentile et al. (2001), who discuss the Central and Southern Florida Project Comprehensive Review Study, use a risk-based conceptual model as they describe the importance of shifting from command-and-control methods to risk-based environmental management.

It is not always easy to build models. Walters (1997) has pointed out some of the difficulties in the management of riparian ecosystems. He believes that management of these systems is hampered by difficulty in modeling and data problems. He cites modeling problems which occur when people try to draw biological conclusions using inappropriate spatial scales, as well as concerns that arise due to the non-additivity of some parameters in population dynamics, among other difficulties.

Even when models are difficult to build, though, adaptive management can be used to help generate the missing knowledge. Two papers (James et al. 2003; Lowry et al. 2003) discuss "gaps" in water management that can be filled by adaptive management. James et al. suggest that adaptive management can be used to fill gaps in water resource managers' knowledge of the effects of salinity on freshwater biota, which would help to construct a stronger model of the ecosystem. Lowry et al. analyze groundwater management in New Zealand and go even further – they suggest that the adaptive management process can be used to fill not only gaps in information about fundamental processes, but also gaps in implementation and management tools.

3. A Range of Management Choices

Walters and Hilborn (1978) discuss adaptive management in the context of optimizing harvesting policies for exploited populations. As an early work in the field, the paper uses optimization models that assume full knowledge of a system, but it also discusses adaptive optimization models that assume very little knowledge, as might be the case in practice. Importantly, the authors emphasize that active adaptive management is a preferable approach to passive adaptive management: one of the key points of the paper is that one needs to generate and test a range of choices in order to find the optimal management strategy.

Bearlin et al. (2002) and van Damme et al. (2003) both focus on modeling as an important tool for selecting the proper policy to implement, but they have also generated multiple management options to model and evaluate. Allison et al. (2004) also model and evaluate multiple options as they come to a conclusion about deactivation of different tracts of logging roads in order to manage landslide risk and ecological health. In another paper, Treves and Karanth (2003) suggest that adaptive management is a key part of controlling the interaction between humans and carnivores, and they discuss various strategies for management of human-carnivore interaction, including eradication, regulated harvest, preservation, modifying behavior, and avoiding the intersection of human and carnivore activities. In these and other cases, generation of management options may also involve stakeholder participation.

Importantly, testing a range of options in an active adaptive management approach, especially if implemented in concert with a control, roughly approximates the scientific method. Conditions in the managed ecosystem will not be as controlled as conditions in a scientific laboratory, but several authors have nonetheless suggested that wildlife management and research be integrated through adaptive management. Lancia et al. (1996), for instance, believe that an adaptive management approach will be a more efficient way to spend monetary resources than separate management and research efforts.

4. Monitoring and Evaluation of Outcomes

Once one has generated and implemented a range of options, monitoring and evaluation are required to determine which option performs the best. Bearlin et al. (2002), van Damme et al. (2003), and Allison et al. (2004), all of whom model the predicted results of their management options, develop project-specific criteria by which to interpret the results. Interestingly, Bearlin et al. (2002) found that the monitoring techniques they modeled were not sufficiently sophisticated to detect differences among management strategy outcomes. While these studies may be helpful for addressing the issue of evaluation, they all took place as modeling exercises rather than in the context of implemented policies.

Many papers have confronted the problem of monitoring a management plan implemented in nature. Hilborn and Sibert (1988) emphasize that careful monitoring is a necessary component of learning to prevent disastrous reductions in fish population due to over-harvesting. Monitoring plays a large role in adaptive management plans set forth by most public agencies; a document developed for the U.S. Army Corps of Engineers (USACE 2003), for example, gives guidance for monitoring within the adaptive management framework. It discusses applications for funding limits, duration of monitoring, and monitoring plans, among other concerns.

A few papers have given general frameworks for monitoring and evaluating adaptive management projects. For the management of multinational large marine ecosystems (LMEs), Duda and Sherman (2002) propose a five-module assessment and management methodology. The modules considered are productivity, fish and fisheries, pollution and ecosystem health, socioeconomics, and governance. For the Nature Conservancy, Parrish et al. (2003) developed a "Measures of Success" framework for assessing the management of protected areas. Their framework has four components: 1) identifying conservation targets, 2) identifying ecological attributes of those targets, 3) identifying acceptable ranges for indicators of those attributes, and 4) rating target status based on the values of the indicators. The results of the assessment process can then feed back into a management loop consisting of objective revisiting, model updating, and option generating.

Some authors have put forth more specific tools for the monitoring process. Hockings (2003) reviews a series of twenty-seven methodologies for assessing the success of protected area management. The methodologies are assessed based on whether they evaluate six criteria – context, planning, input, process, output, and outcome – that should be monitored to evaluate a management plan. Because it can be unclear how to manipulate data once it is collected to evaluate a management plan, Hockings suggests either quantitative monitoring or qualitative evaluation through scoring by stakeholders. To aid adaptive managers in data analysis, Sit and Taylor (1998) developed a statistical methods handbook for adaptive managers. The tools discussed in the book are unique because many standard experimental statistical methods cannot be applied to adaptive management scenarios. The tools deal with the design of experiments, studies of uncontrolled events, retrospective studies, measurements and estimates, errors of inference, Bayesian statistical methods, decision analysis, and selecting the appropriate statistical methods.

5. Mechanisms for Incorporating Learning into Future Decisions

The central idea of adaptive management is to gain knowledge and reduce uncertainty about the system being managed through scientific strategies, including the generation and testing of multiple management options. Effort would be wasted if the process lacked a mechanism for incorporating learning into future decisions; indeed, some studies use a framework to select the present optimal strategy, but they do not feed that information back into the system for future use and strategy revision. Fortunately, there are also many projects which do include an information feedback mechanism.

The best examples of information feedback are found in large, ongoing projects. Imperial et al. (1993) trace the development of the National Estuary

Program (NEP) from two predecessor programs and evaluate the effectiveness of the NEP in estuary management. They note that adaptive management is well-suited to the NEP because lessons learned from past estuary management can be applied to new estuaries becoming part of the program. Similarly, Voss (2000) describes the Central and Southern Florida Project Comprehensive Review Study, a restoration plan for the Florida Everglades entailing sixty water management projects. Voss notes that adaptive management fits a project of this size particularly well because knowledge gained from adaptive management can be used both to modify the management plan for an individual project and to plan future projects.

In addition to leading to revised management plans, learning can assist the modeling phase of adaptive management as well. Fish population numbers in fisheries are typically extremely volatile, and Charles (1998) proposes learning through adaptive management as a way to mitigate the uncertainty inherent in fishery management. As in other application areas, iterations of adaptive management are useful for reducing uncertainty in parameter estimates and structural uncertainty in fishery modeling. Charles gives examples of cases in which adaptive management principles have been used and other cases in which adaptive management could have helped thwart disaster.

In most adaptive management projects, learning is an important goal. The restoration of the Florida Everglades, for example, is a huge challenge with tremendous uncertainty. Kiker et al. (2001) emphasize that the current knowledge base is insufficient for Everglades restoration, and they advance adaptive management as a way to supplement scientific learning and promote holistic understanding useful to decision-makers and political representatives. They are by no means the only ones to espouse learning as an objective: in the context of Canadian hydroelectric power and fisheries, McDaniels and Gregory (2004) recommend that learning be a goal of the management decision process; in the context of achieving biodiversity, Salafsky et al. (2002) emphasize the development of knowledge and conservation skills through adaptive management; and in the context of Chilean fisheries, Castilla (2000) proposes a plan to use adaptive management and other tools in the establishment of research programs and fishery sites. All of these plans use adaptive management as a way to better understand the managed ecosystem.

6. A Collaborative Structure for Stakeholder Participation and Learning

Adaptive management is not conducted by managers or scientists alone: an important element of adaptive management is the inclusion of stakeholders, allowing them both to give input to the adaptive management process and to gain knowledge from it. When Hillman and Brierly (2002) assess the information needed to adaptively manage the Lachlan River in New South Wales, Australia, the information they identify includes socioeconomic and political factors in addition to scientific ones. Sutinen (2000) provides a step-by-step framework for the monitoring and assessment of large marine ecosystems, including the evaluation of socioeconomics and stakeholder input. Moreover, updating managerial objectives and strategies through active adaptive management involves admitting the uncertainty inherent in the system being managed. Unless all stakeholders are involved in the learning and decision-making processes, the uncertainty can be hard for the manager to admit. Collaboration is thus essential to a successful adaptive management plan: in order for management policies to be acceptable to everyone in the community, all stakeholders need to be involved in the adaptive management process, especially its learning aspects.

Many case studies have shown that public involvement is essential to adaptive management. Pinkerton (1999) believes that when adaptive management fails to incorporate the general public and agencies into decisions, distrust and political tension result. Gunderson (1999) looks at the Florida Everglades and notes the inflexibility of stakeholders as a primary impediment to adaptive management, while Gilmour et al. (1999) highlight the usefulness of adaptive management workshops in promoting understanding between participants and managers. Shindler and Aldred Cheek (1999) likewise conclude that public involvement in management decisions is essential to successful adaptive management implementation. After briefly reviewing these and other studies, Johnson (1999) found integrating stakeholders into decision-making and developing institutions that are amenable to adaptive management were key elements of the process.

Folke (2003) introduces the concept of adaptive co-management, which is essentially adaptive management with multiple groups as decision-makers. Folke suggests that using co-management can be an easier way to deal with unexpected results of a management strategy because it minimizes stakeholder objections after a decision has been made. But the benefits of collaboration are not limited to public acceptance of the management strategy: Lee (1999) believes that all aspects of the adaptive management process can be improved through more collaboration among stakeholders.

INTEGRATION OF ADAPTIVE MANAGEMENT WITH MULTI-CRITERIA DECISION ANALYSIS

When faced with significant uncertainty and the inability to predict the dynamics of natural ecosystems, scientists in the field of natural resource management introduced the concept of adaptive management. Unlike traditional management schemes designed to find and follow the optimal remedial strategy, the adaptive management paradigm explicitly acknowledges the high degree of uncertainty in system dynamics and our inability to predict system evolution in response to changing physical environments and social pressures. Our review indicates that the concept of adaptive management is well respected in academia, and many government agencies have recommended it for application. Most of the applications, though, are limited to large-scale natural resource management (such as the Everglades and Grand Canyon National Park). Even though many papers acknowledge that effective environmental decision-making requires considering the environmental, ecological, technological, economic, and socio-political factors relevant to evaluating and selecting a management alternative, these factors are rarely considered in concert and decisions are often driven by just one aspect of the problem. The quantitative tools and methods for implementing adaptive management strategies are not systematized, and no framework is available for integrating and organizing the people, processes, and tools required to make structured and defensible environmental management decisions.

Realization of our inability to model and predict ecosystem response, as well as a regulatory push for transparent decision-making processes and explicit trade-offs, is currently driving adaptive management from its traditional niche in natural resource management to other areas. Large-scale modeling comparisons have revealed a large degree of uncertainty in model predictions even for simple ecosystems. For example, Linkov and Burmistrov (2003) report a difference of up to seven orders of magnitude in predicting future radionuclide concentrations by several regulator-approved models in a strawberry plant sprayed with contaminants under well-controlled conditions. In addition to this degree of uncertainty in model prediction, dissatisfactory results of long-term monitoring at Superfund and other sites with optimization-based management schemes have resulted in regulatory attempts to improve the system by adding optimization loops after a decision is made (ITRC 2004; EPA 2000).

In our previous papers (Linkov et al. 2004; Kiker et al. 2005; Linkov et al. 2005; Linkov et al. 2006a) we proposed multi-criteria decision analysis (MCDA) as a tool for integrating heterogeneous information (technical, social, political) as well as for the explicit incorporation of decision-makers and stakeholder (including public) value judgments. MCDA begins with a collectivity of decision-makers, scientists, and other stakeholders defining a problem and generating alternatives to solve that problem. The group then decides what criteria it will use to judge the alternatives against one another. The alternatives are judged along each of the criteria, the relative importance of each of the criteria is determined, and the criteria scores are systematically compared to choose the best alternative. MCDA is a structured decision-making process that begins with problem formulation and ends with the selection of an alternative to solve that problem. The advantages of using MCDA techniques over other less structured decision-making methods are numerous. MCDA provides a clear and transparent way of making decisions and provides a formal way to combine information from disparate sources. These qualities make decisions made using MCDA more defendable than decisions made in less structured manners.

We believe that a combination of adaptive management and MCDA will provide a powerful framework for a wide range of environmental management problems. It will allow for both structured, clear decisions and the adjustment of those decisions based on their performance. The MCDA framework proposed in Linkov et al. (2004) and in many other papers incorporates feedback loops between each step of the process that leads to the selection of an alternative. Adoption of adaptive management views will allow for the iterative application of the MCDA framework. Once an alternative is chosen, reintroducing a feedback loop will allow the re-ranking of alternatives as well as goals and criteria weightings.

A systematic decision framework (Figure 3.3) is intended to provide a generalized road map to the environmental decision-making process. Having the right combination of people is the first essential element in the decision process. The activity and involvement levels of three basic groups of people (decision-makers, scientists and engineers, and stakeholders) are symbolized in Figure 3.3 by dark lines for direct involvement and dotted lines for less direct involvement. While the actual membership and the functions of these three groups may overlap or vary, the roles of each are essential in maximizing the utility of human input into the decision process. Each group has its own way of viewing the world, its own method of envisioning solutions, and its own societal responsibility. Policy- and decision-makers spend most of their effort defining the problem's context and the overall constraints on the decision. In addition, they may have responsibility for final policy selection and implementation. Stakeholders may provide input in defining the problem, but contribute the most input into helping formulate performance criteria and making value judgments for weighting the various success criteria. Depending on the problem and regulatory context, stakeholders may have some responsibility in ranking and selecting the final option. Scientists and engineers have the most focused role in that they provide the measurements or estimations of the desired criteria that determine the success of various alternatives. While they may take a secondary role as stakeholders or decision-makers, their primary role is to provide the technical input necessary for the decision process.

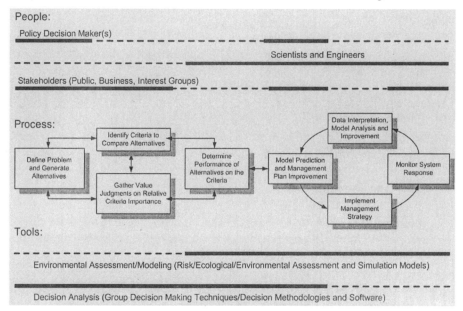

Figure 3.3. Framework for environmental decision-making process.

The decision-making process is in the center of the figure. While it is reasonable to expect that the process may vary in specific details among regulatory programs and project types, emphasis should be given to designing an adaptive management structure so that participants can modify aspects of the project to suit local concerns while still producing a structure that provides the required outputs. The process depicted in Figure 3.3 follows two basic themes: 1) generating management alternatives, success criteria, and value judgments and 2) ranking the alternatives by applying value weights. The first part of the process generates and defines choices, performance levels, and preferences. The latter section methodically prunes non-feasible alternatives by first applying screening mechanisms (for example, overall cost, technical feasibility, or general societal acceptance) followed by a more detailed ranking of the remaining options by decision analytical techniques (analytical hierarchy process, multi-attribute utility theory, or outranking) that utilize the various criteria levels generated by environmental tools, monitoring, or stakeholder surveys.

As shown in Figure 3.3, the tools used within group decision-making and scientific research are essential elements of the overall decision process. As with people, the applicability of the tools is symbolized by solid lines (direct or high utility) and dotted lines (indirect or lower utility). Decision analysis tools help to generate and map stakeholder preferences as well as individual value judgments into organized structures that can be linked with other technical tools from risk analysis, modeling and monitoring, and cost estimation. Decision analysis software can also provide useful graphical techniques and visualization methods to express the gathered information in understandable formats. When changes occur in the requirements or the decision process, decision analysis tools can respond efficiently to reprocess and iterate with the new inputs. The framework depicted in Figure 3.3 provides a focused role for the detailed scientific and engineering efforts invested in experimentation, environmental monitoring, and modeling that provide the rigorous and defendable details for evaluating criteria performance under various alternatives. This integration of decision tools and scientific and engineering tools allows each to have a unique and valuable role in the decision process without attempting to apply either type of tool beyond its intended scope.

As with most other decision processes reviewed, it is assumed that the framework in Figure 3.3 is iterative at each phase and can be cycled through many times in the course of complex decision-making. A first-pass effort may efficiently point out challenges that may occur, key stakeholders to be included, or modeling studies that should be initiated. As these challenges become more apparent, one iterates again through the framework to explore and adapt the process to address the more subtle aspects of the decision, with each iteration giving an indication of additional details that would benefit the overall decision.

We are not the first to propose an integration of decision analysis methods with adaptive management. For example, Rauscher et al. (2000) propose a combination of multi-criteria decision analysis and adaptive management. They use an MCDA technique called the analytical hierarchy process (AHP) while proposing that the chosen forest management strategy is monitored, an integral element of adaptive management. Pastorok et al. (1997) also develop a decision-making framework combined with monitoring and adaptive management. Additionally, Nudds et al. (2003), at the University of Guelph Decision Analysis and Adaptive Management Project, are applying a combination of both methods to the management of yellow perch and walleye in Lake Erie. Their system emphasizes decision analysis, placing adaptive management in its service as a method to reduce the uncertainty present in decision calculations. Our proposed method, on the other hand, uses adaptive management as an overall planning and procedural framework, and we recommend decision analysis to help make structured, logical decisions concerning management options. We believe that our method will prove most fruitful, and we believe that our current review provides the most rigorous and justifiable foundation to date for successful application of the combined concept for environmental management of a wide range of projects.

REFERENCES

Allison, C., R. C. Sidle, and D. Tait. 2004. "Application of Decision Analysis to Forest Road Deactivation in Unstable Terrain." *Environmental Management* 33: 173-185.

Bearlin, A. R., E. S. G. Schreiber, S. J. Nicol, A. M. Starfield, and C. R. Todd. 2002. "Identifying the Weakest Link: Simulating Adaptive Management of the Reintroduction of a Threatened Fish." *Canadian Journal of Fish and Aquatic Science* 59: 1709-1716.

Bliss, J. C., S. B. Jones, J. A. Stanturf, M. K. Burke, and C. M. Hamner. 1997. "Creating a Knowledge Base for Management of Southern Bottomland Hardwood Ecosystems." USDA Forest Service, Southern Research Station, General Technical Report SRS-017. http://www.srs.fs.usda.gov/pubs/ viewpub.jsp?index=2203. Accessed 22 July 2004.

British Columbia Forest Service (BCFS). 2000. http://www.for.gov.bc.ca/hfp/amhome/ am_case_studies.htm. Accessed 17 September 2004.

British Columbia Ministry of Sustainable Resource Management (BCMSRM). 2004. North Coast Land and Resource Management Plan: Final Recommendations. http://srmwww.gov.bc.ca/ske/lrmp/ncoast/docs/NCLRMP_Recommendations_Draft_ 4.pdf. Accessed 28 September 2004.

Bureau of Land Management (BLM). 2003. http://www.mt.blm.gov/oilgas/policy/ im2003-234.pdf. Accessed 28 September 2004.

Bunch, M. J., and D. J. Dudycha. 2004. "Linking Conceptual and Simulation Models of the Cooum River: Collaborative Development of a GIS-Based DSS for Environmental Management." *Computers, Environment and Urban Systems* 28: 247-264.

Bundy, A. 2004. "The Ecological Effects of Fishing and Implications for Coastal Management in San Miguel Bay, the Philippines." *Coastal Management* 32: 25-38.

Castilla, J. C. 2000. "Roles of Experimental Marine Ecology in Coastal Management and Conservation." *Journal of Experimental Marine Biology and Ecology* 250: 3-21.

CCHO, Health Canada. 2000. *Climate Change and Health Economic Advisory Panel's Final Report on Health Impacts of the Greenhouse Gases Mitigation Measures.* http://www.hc-sc.gc.ca/hecs-sesc/ccho/publications/greenhouse_gases/toc.htm. Accessed 17 September 2004.

Commission of the European Communities (CEC). 2001. *Green Paper on the Future of the Common Fisheries Policy.* http://europa.eu.int/eur-lex/en/com/gpr/2001/ com2001_0135en01.pdf. Accessed 17 September 2004.

Charles, A. T. 1998. "Living with Uncertainty in Fisheries: Analytical Methods, Management Priorities and the Canadian Groundfishery Experience." *Fisheries Research* 37: 37-50.

Crumbling, D., J. Hayworth, B. Call, W. Davis, R. Howe, D. Miller, and R. Johnson. 2004. "The Maturing of the Triad Approach: Avoiding Misconceptions." *Remediation: The Journal of Environmental Cleanup Costs, Technologies, & Techniques* 14 (4): 81-96.

Dallmeier, F., A. Alonso, and M. Jones. 2002. "Planning an Adaptive Management Process for Biodiversity Conservation and Resource Development in the Camisea River Basin." *Environmental Monitoring and Assessment* 76 (1): 1-17.

DOE. 1996. *Comprehensive Land-Use Planning Process Guide.* http://www.sc.doe.gov/SC-80/pdf_file/gpg33.pdf. Accessed 20 September 2004.

DOE. 2002. *Adaptive Management Platform for Natural Resources in the Columbia River Basin.* http://www.pnl.gov/main/publications/external/technical_reports/PNNL-13875.pdf. Accessed 28 September 2004.

DOE. 2003. *The NEPA Task Force Report to the Council on Environmental Quality: Modernizing NEPA Implementation.* http://ceq.eh.doe.gov/ntf/report/finalreport.pdf. Accessed 20 September 2004.

DOI. 2004. *Departmental Manual*. http://elips.doi.gov/elips/release/3614.htm and http://elips.doi.gov/elips/release/3611.htm. Accessed 29 September 2004.

Duda, A., and K. Sherman. 2002. "A New Imperative for Improving Management of Large Marine Ecosystems." *Ocean & Coastal Management* 45: 797-833.

Environment Canada. 2004a. http://www.bco.ec.gc.ca/en/primers/ecosystem.cfm. Accessed 17 September 2004.

Environment Canada. 2004b. *Environmental Assessment Best Practice Guide for Wildlife at Risk in Canada*. http://www.cwsscf.ec.gc.ca/publications/ AbstractTemplate.cfm?lang=e&id=1059. Accessed 17 September 2004.

Environment Canada. 2004c. *Canada's National Wildlife Disease Strategy*. http://www.cws-scf.ec.gc.ca/cnwds/draft11.pdf. Accessed 20 September 2004.

EPA. 1999a. *Integrated Assessment of Hypoxia in the Northern Gulf of Mexico*. http://www.epa.gov/msbasin/ia/index.html. Accessed 17 September 2004.

EPA. 1999b. *A Sustainable Brownfields Model Framework*. http://www.epa.gov/brownfields/pdf/susmodel.pdf. Accessed 23 September 2004.

EPA. 1999c. *Protocol for Developing Sediment TMDLs*. http://www.epa.gov/owow/tmdl/sediment/pdf/sediment.pdf. Accessed 23 September 2004.

EPA. 2000. *Watershed Analysis and Management Guide for Tribes – Step 5: Adaptive Management*. http://www.epa.gov/owow/watershed/wacademy/wam/step5.html. Accessed 17 September 2004.

EPA. 2001a. *Protocol for Developing Pathogen TMDLs*. http://www.epa.gov/owow/tmdl/pathogen_all.pdf. Accessed 23 September 2004.

EPA. 2001b. *Action Plan for Reducing, Mitigating, and Controlling Hypoxia in the Northern Gulf of Mexico*. http://www.epa.gov/msbasin/actionplan.htm. Accessed 29 September 2004.

EPA. 2002. *Lake Superior Lakewide Management Plan (LaMP) 2000*. http://www.epa.gov/glnpo/lakesuperior/lamp2000/. Accessed 17 September 2004.

EPA. 2003. *Using Dynamic Field Activities for On-Site Decision Making: A Guide for Project Managers*. http://www.epa.gov/superfund/programs/dfa/download/ guidance/40r03002.pdf. Accessed 3 February 2006.

EPA. 2004a. *Guidance for Monitoring at Hazardous Waste Sites: Framework for Monitoring Plan Development and Implementation*. http://www.epa.gov/superfund/action/guidance/dir9355.pdf. Accessed 23 September 2004.

EPA. 2004b. *Terminology Reference System: Adaptive Management*. http://oaspub.epa.gov/trs/trs_proc_qry.navigate_term?p_term_id=8882&p_term_cd= TERM. Accessed 20 September 2004.

EPA. 2005. *Welcome to Superfund*. http://www.epa.gov/superfund/. Accessed 30 November, 2005.

FERC. 2000. *Interagency Task Force Report on Improving the Studies Process in FERC Licensing*. http://www.nmfs.noaa.gov/habitat/habitatprotection/ studies_final.pdf. Accessed 20 September 2004.

Folke, C. 2003. "Freshwater for Resilience: A Shift in Thinking." *Philosophical Transactions of the Royal Society of London*. 358: 2027-2036.

GEF. 2004. http://www.gefweb.org/MonitoringandEvaluation/MEAbout/meabout.html. Accessed 17 September 2004.

Gentile, J. H., M. A. Harwell, W. Cropper, Jr., C. C. Harwell, D. DeAngelis, S. Davis, J. C. Ogden, and D. Lirman. 2001. "Ecological Conceptual Models: A Framework and Case Study on Ecosystem Management for South Florida Sustainability." *The Science of the Total Environment*, 274: 231-253.

Gilmour, A., G. Walkerden, and J.Scandol. 1999. "Adaptive Management of the Water Cycle on the Urban Fringe: Three Australian Case Studies." *Conservation Ecology* 3 (1): 11. http://www.consecol.org/vol3/iss1/art11. Accessed 14 July 2004.

Gunderson, L. 1999. "Resilience, Flexibility and Adaptive Management -- Antidotes for Spurious Certitude?" *Conservation Ecology* 3 (1): 7. http://www.consecol.org/vol3/iss1/art7. Accessed 14 July 2004.

Habron, G. 2003. "Role of Adaptive Management for Watershed Councils." *Environmental Management* 3 (11): 29-41.

Haney, A., and R. L. Power. 1996. "Adaptive Management for Sound Ecosystem Management." *Environmental Management* 20 (6): 879-886.

Hilborn, R., and J. Sibert. 1988. "Adaptive Management of Developing Fisheries." *Marine Policy* 12: 112-121.

Hillman, M., and G. Brierly. 2002. "Information Needs for Environmental Flow Allocation: A Case Study from the Lachlan River, New South Wales, Australia." *Annals of the Association of American Geographers* 92: 617-630.

Hockings, M. 2003. "Systems for Assessing the Effectiveness of Management in Protected Areas." *Bioscience* 53: 823-832.

Imperial, M. B., T. Hennessey, and D. Robadue, Jr. 1993. "The Evolution of Adaptive Management for Estuarine Ecosystems: The National Estuary Program and Its Precursors." *Ocean & Coastal Management* 20: 147-180.

Innes, J., R. Hay, I. Flux, P. Bradfield, H. Speed, and P. Jansen. 1999. "Successful recovery of North Island Kokako Callaeas Cinerea Wilsoni Populations, by Adaptive Management." *Biological Conservation* 87 (2): 201-214.

ITRC. 2004. *Remediation Process Optimization: Identifying Opportunities for Enhanced and More Efficient Site Remediation.* http://www.itrcweb.org/RPO%20-1_edit2.pdf. Accessed 23 September 2004.

James, K. R., B. Cant, and T. Ryan. 2003. "Responses of Freshwater Biota to Rising Salinity Levels and Implications for Saline Water Management: A Review." *Australian Journal of Botany* 51: 703-713.

Johnson, B. L. 1999. "Introduction to the Special Feature: Adaptive Management - Scientifically Sound, Socially Challenged?" *Conservation Ecology* 3 (1): 10. http://www.consecol.org/vol3/iss1/art10. Accessed 14 July 2004.

Johnson, F., and K. Williams. 1999. "Protocol and Practice in the Adaptive Management of Waterfowl Harvests." *Conservation Ecology* 3 (1): 8. http://www.consecol.org/vol3/iss1/art8. Accessed 14 July 2004.

Kiker, C. F., J. W. Milon, and A. W. Hodges. 2001. "Adaptive Learning for Science-Based Policy: The Everglades Restoration." *Ecological Economics* 37: 403-416.

Kiker, G., T. Bridges, I. Linkov, A. S. Varghese, and T. P. Seager. 2005. "Application of Multi-Criteria Decision Analysis in Environmental Management." *Integrated Environmental Assessment and Management* 2: 1-14.

Lancia, R. A., C. E. Braun, M. W. Collopy, and R. D. Dueser. 1996. "ARM! for the Future: Adaptive Resource Management in the Wildlife Profession." *Wildlife Society Bulletin* 24: 436-442.

Lee, K. N. 1999. "Appraising Adaptive Management. *Conservation Ecology* 3 (2): 3. http://www.consecol.org/vol3/iss2/art3. Accessed 15 July 2004.

Lempert, R., S. Popper, and S. C. Bankes. 2003. "Shaping the Next One Hundred Years: New Methods for Quantitative, Long-Term Policy Analysis." http://www.rand.org/publications/MR/MR1626/. Accessed 28 September 2004.

Linkov, I., and D. Burmistrov. 2003. "Model Uncertainty and Choices Made by Modelers: Lessons Learned from the International Atomic Energy Agency Model Intercomparisons." *Risk Analysis* 23 (6): 1297-1308.

Linkov, I., A. Varghese, S. Jamil, T. P. Seager, G. Kiker, and T. Bridges. 2004.

"Multi-Criteria Decision Analysis: Framework for Applications in Remedial Planning for Contaminated Sites." In *Comparative Risk Assessment and Environmental Decision Making,* I. Linkov and A. Ramadan, eds., 15-54. Amsterdam: Kluwer.

Linkov, I., S. Sahay, T. P. Seager, G. Kiker, and T. Bridges. 2005. "Multi-Criteria Decision Analysis: A Framework for Managing Contaminated Sediments." In *Strategic Management of Marine Ecosystems,* E. Levner, I. Linkov, and J. M. Proth, eds., 271-297. Amsterdam: Springer.

Linkov, I., S. Sahay, T. Seager, G. Kiker, T. Bridges, D. Belluck, and A. Meyer. 2006a. "Multi-Criteria Decision Analysis: Comprehensive Decision Analysis Tool for Risk Management of Contaminated Sediments." *Risk Analysis* 26: 1-18.

Linkov, I., K. Satterstrom, G. Kiker, T. Bridges, S. Benjamin, and D. Belluck. 2006b. "From Optimization to Adaptation: Shifting Paradigms in Environmental Management and Their Application to Remedial Decisions." *Integrated Environmental Assessment and Management* 2: 92-98.

Lowry, T. S., J. C. Bright, M. E. Close, C. A. Robb, P. A. White, and S. G. Cameron. 2003. "Management Gaps Analysis: A Case Study of Groundwater Resource Management in New Zealand." *Water Resources Development* 19: 579-592.

McDaniels, T. L., and R. Gregory. 2004. "Learning as an Objective within a Structured Risk Management Decision Process." *Environmental Science and Technology* 38: 1921-1926.

McGinley, K., and B. Finegan. 2003. "The Ecological Sustainability of Tropical Forest Management: Evaluation of the National Forest Management Standards of Costa Rica and Nicaragua, with Emphasis on the Need for Adaptive Management." *Forest Policy and Economics* 5 (4): 421-431.

McNeeley, J. A. 2004. "Nature vs. Nurture: Managing Relationships between Forests, Agroforestry and Wild Biodiversity." *Agroforestry Systems* 61:155-165.

Marmorek, D., and C. Peters. 2001. "Finding a PATH toward Scientific Collaboration: Insights from the Columbia River Basin." *Conservation Ecology* 5 (2): 8. http://www.consecol.org/vol5/iss2/art8. Accessed 9 February 2005.

Murray, C., and D. Marmorek. 2003. "Adaptive Management and Ecological Restoration." In *Ecological Restoration of Southwestern Ponderosa Pine Forests: A Sourcebook for Research and Application,* P. Friederici, ed., 417-428. Washington, DC: Island Press.

NOAA. 2004a. http://www.csc.noaa.gov/coastal/management/management.htm. Accessed 17 September 2004.

NOAA. 2004b. http://www.csc.noaa.gov/coastal/management/monitor.htm. Accessed 17 September 2004.

NOAA. 2004c. http://www.csc.noaa.gov/lcr/text/confsumm.html. Accessed 29 September 2004.

Norton, B. G., and A. C. Steinemann. 2001. "Environmental Values and Adaptive Management.*Environmental Values* 10 (4): 473-506.

NPS. 2002. http://www.nps.gov/grte/winteruse/seis/table11.pdf. Accessed 28 September 2004.

NRC (National Research Council) Committee on Environmental Remediation at Naval Facilities, Water Science and Technology Board, Division on Earth and Life Studies. 2003. *Environmental Cleanup at Navy Facilities.* Washington, DC: National Academies Press.

NRC (National Research Council). Panel on Adaptive Management for Resource Stewardship, Committee to Assess the U.S. Army Corps of Engineers Methods of Analysis and Peer Review for Water Resources Project Planning, Water Science and Technology Board, Ocean Studies Board, Division on Earth and Life Studies. 2004. *Adaptive Management for Water Resources Planning.* Washington, DC: National Academies Press.

Nudds, T., Y. Jiao, S. Crawford, K. Reid, K. McCann, and W. Yang. 2003. *The DAAM Project – Decision Analysis and Adaptive Management (DAAM) Systems for Great Lakes Fisheries: The Lake Erie Walleye and Yellow Perch Fisheries Project Background and Work Plan.* http://www.ocfa.on.ca/PICS/DAAM_backgrounder_2003_11_26.pdf. Accessed 9 February 2005.

Nyberg, B. 1999. *An Introductory Guide to Adaptive Management for Project Leaders and Participants.* Forest Practices Branch, British Columbia Forest Service.

Parrish, J. D., D. P. Braun, and R. S. Unnasch. 2003. "Are We Conserving What We Say We Are? Measuring Ecological Integrity Within Protected Areas." *Bioscience* 53: 851-860.

Pastorok, R. A., A. MacDonald, J. R. Sampson, and P. Wilber. 1997. "An Ecological Decision Framework for Environmental Restoration Projects." *Ecological Engineering* 9: 89-107.

Pinkerton, E. 1999. "Factors in Overcoming Barriers to Implementing Co-Management in British Columbia Salmon Fisheries." *Conservation Ecology* 3 (2): 2. http://www.consecol.org/vol3/iss2/art2. Accessed 15 July 2004.

Rauscher, H. M. 1999. "Ecosystem Management Decision Support for Federal Forests in the United States: A Review." *Forest Ecology and Management* 114: 173-197.

Rauscher, H. M., F. T. Lloyd, D. L. Loftis, and M. J. Twery. 2000. "A Practical Decision-Analysis Process for Forest Ecosystem Management." *Computers and Electronics in Agriculture* 27: 195-226.

Rudd, M. A. 2004. "An Institutional Framework for Designing and Monitoring Ecosystem-Based Fisheries Management Policy Experiments." *Ecological Economics* 48:109-124.

Salafsky, N., R. Margoluis, K. H. Redford, and J. G. Robinson. 2002. "Improving the Practice of Conservation: A Conceptual Framework and Research Agenda for Conservation Science." *Conservation Biology* 16 (6): 1469-1479.

Southeastern Adaptive Management Group (SEAMG). 2004. http://cars.er.usgs.gov/SEAMG/seamg.html. Accessed 28 September 2004.

Shindler, B., and K. Aldred Cheek. 1999. "Integrating Citizens in Adaptive Management: A Propositional Analysis." *Conservation Ecology* 3 (1): 9. http://www.consecol.org/vol3/iss1/art9. Accessed 14 July 2004.

Sit, V., and B. Taylor, eds. 1998. *Statistical Methods for Adaptive Management Studies.* Land Management Handbook. British Columbia Ministry of Forests Research Program.

Steyer, G. D., and D. W. Llewellyn. 2000. "Coastal Wetlands Planning, Protection, and Restoration Act: A Programmatic Application of Adaptive Management." *Ecological Engineering* 15: 385-395.

Sutinen, J. G., ed. 2000. *A Framework for Monitoring and Assessing Socioeconomics and Governance of Large Marine Ecosystems.* NOAA Technical Memorandum NMFS-NE-158.

Szaro, R. C., J. Berc, S. Cameron, S. Cordle, M. Crosby, L. Martin, D. Norton, R. O'Malley, and G. Ruark. 1998. "The Ecosystem Approach: Science and Information Management Issues, Gaps and Needs." *Landscape and Urban Planning* 40 (1-3): 89-101.

Taylor, B., L. Kremsater, and R. Ellis. 1997. *Adaptive Management of Forests in British Columbia.* British Columbia Ministry of Forests, Forest Practices Branch. http://www.for.gov.bc.ca/hfp/amhome/am_publications.htm. Accessed 20 July 2004.

Thom, R. H. 1997. "System-Development Matrix for Adaptive Management of Coastal Ecosystem Restoration Projects." *Ecological Engineering* 8: 219-232.

Thom, R. H. 2000. "Adaptive Management of Coastal Ecosystem Restoration Projects." *Ecological Engineering* 15: 365-372.

Treves, A., and K. U. Karanth. 2003. "Human-Carnivore Conflict and Perspectives on Carnivore Management Worldwide." *Conservation Biology* 17 (6): 1491-1499.

U.S. Army Corps of Engineers (USACE). 2003. *Monitoring and Adaptive Management.* Planning Associates Program, United States Army Corps of Engineers.

USBR. 2001. Draft Battle Creek Salmon and Steelhead Restoration Project Adaptive Management Plan. http://www.usbr.gov/mp/regional/battlecreek/docs/AMP_PublicReview-clean.pdf. Accessed 28 September 2004.

USDA. 2003a. *Ecological System Principles.* http://policy.nrcs.usda.gov/scripts/lpsiis.dll/H/H_180_600_D_44.htm. Accessed 20 September 2004.

USDA. 2003b. *Monitoring.* http://policy.nrcs.usda.gov/scripts/lpsiis.dll/H/H_190_610_B_35.htm. Accessed 20 September 2004.

USGS. 2004. *Developing an Adaptive Management Approach Using Offsets for Reducing Mercury Loadings to the Sacramento River Watershed.* http://geography.wr.usgs.gov/science/mercury-tmdl.html. Accessed 20 September 2004.

van Damme, L., J. S. Russell, F. Doyon, P. N. Duinker, T. Gooding, K. Hirsch, R. Rothwell, and A. Rudy. 2003. "The Development and Application of a Decision Support System for Sustainable Forest Management on the Boreal Plain." *Journal of Environmental Engineering Science* 2: S23-S24.

van Staden, V., B. F. N. Erasmus, J. Roux, M. J. Wingfield, and A. S. van Jaarsveld. 2004. "Modelling the Spatial Distribution of Two Important South African Plantation Forestry Pathogens." *Forest Ecology and Management* 187: 61-73.

Voss, M. 1999. "The Central and Southern Florida Project Comprehensive Review Study: Restoring the Everglades." *Ecology Law Quarterly* 27: 751-770.

Walters, C. 1997. "Challenges in Adaptive Management of Riparian and Coastal Ecosystems." *Conservation Ecology* 1 (2): 1. http://www.consecol.org/vol1/iss2/art1. Accessed 14 July 2004.

Walters, C. J., and R. Hilborn. 1978. "Ecological Optimization and Adaptive Management." *Annual Review of Ecological Systems.* 9: 157-188.

Weinstein, M. P., J. H. Balletto, J. M. Teal, and D. F. Ludwig. 1997. "Success Criteria and Adaptive Management for a Large-Scale Wetland Restoration Project." *Wetlands Ecology and Management* 4: 111-127.

Weishar, L. L., and J. M. Teal. 1998. "The Role of Adaptive Management in the Restoration of Degraded Diked Salt Hay Farm Wetlands." Denver, CO: Proceedings of the ASCE Wetlands Engineering & River Restoration Conference.

Wilhere, G. F. 2002. "Adaptive Management in Habitat Conservation Plans." *Conservation Biology* 16: 20-29.

World Wildlife Federation (WWF). 2001. "Put Environment at the Heart of European Fisheries Policy: WWF Manifesto for the Review of the EU Common Fisheries Policy." http://www.panda.org/downloads/marine/manifesto_1.pdf. Accessed 17 September 2004.

World Wildlife Federation (WWF). 2001. "Adaptive Management: A Tool for Conservation Practitioners." http://effectivempa.noaa.gov/docs/adaptive.pdf. Accessed 20 September 2004.

Zedler, J. B. 1997. "Adaptive Management of Coastal Ecosystems Designed to Support Endangered Species." *Ecology Law Quarterly* 24(4): 735-743.

Reclaiming the Land

APPENDIX

Table 3.1. Government application and guidance documents.

Agency	Topic	Methods / Tools	Application Area	1	2	3	4	5	6	Citation(s)
Environmental Protection Agency	Total maximum daily loads (TMDLs) for pollutants in the environment	Follow-up monitoring and evaluation	TMDLs for sediments, pathogens				■	·		EPA, 1999c; EPA, 2001a
	Watershed Management Guide for Tribes	Implementation monitoring, validation monitoring, effectiveness monitoring	Watershed management				■	·		EPA, 2000
	Management of contaminated or hazardous sites, including brownfields	Sustainable development, monitoring framework	Hazardous site management	·			■		·	EPA, 1999b; EPA, 2004c
	Integrated Assessment of Hypoxia in the northern Gulf of Mexico	Models to interpret change in hydrologic and ecological systems	Coastal and marine ecosystems		■		■	·		EPA, 1999a; EPA, 2001b
	Lake Superior Lakewide Management Plan (LaMP), 2000	Extensive stakeholder involvement in LaMPs	Lake management	·			■	■	■	EPA, 2002a
	The "Triad" methodology for hazardous site management, applied specifically to brownfields	Systematic project planning, dynamic work plan strategies, and real-time measurement in service of a pre-fabricated decision support matrix	Hazardous site management		■	·	■	■	·	EPA, 2003
Department of the Interior	DOI's departmental manual	Adaptive management required of all bureau heads	NEPA compliance			·				DOI, 2004
	Southeastern Adaptive Management Group, better integration between research and management	Ecological and statistical theory, analytical and decision-support tools, and institutional arrangements	Adaptive management group				■		■	SEAMG, 2004

Elements of AM

Agency	Topic	Methods / Tools	Application Area	Elements of AM 1	2	3	4	5	6	Citation(s)
	The Bureau of Land Management's land use authorizations for oil and gas programs	Emphasis on monitoring	Land use	·			·			BLM, 2003
	Restoration and enhancement of 42 miles of fish habitat in California while simultaneously supporting hydroelectricity generation	Detailed monitoring	Watershed management		·			·	■	USBR, 2001
	Management of mercury contamination of Sacramento River Watershed	Offsets, modeling, monitoring	TMDLs, watershed management	■		·	■		■	USGS, 2004
National Marine Fisheries Service	Socioeconomic and governance-related human dimensions of managing large marine ecosystems	Interaction matrices	Coastal and marine ecosystems				■		■	Sutinen, 2000
	Coastal restoration and management guidance; case study in Rhode Island marsh restoration	Well-formulated objectives and detailed monitoring, system-development matrix	Coastal and marine ecosystems		·		■	·	■	NOAA, 2004a; NOAA, 2004b; NOAA, 2004c
Department of Energy	Land use planning and process framework	Learning through monitoring	Land use				■	■		DOE, 1996
	Modernizing NEPA Implementation	Learning through monitoring	NEPA				■	·		DOE, 2003
	Adaptive management of salmon and hydroelectricity in the Columbia River basin	Models and computing power	Hydropower	■	■	·	■	■	■	DOE, 2002
	Adaptive management of hydroelectricity in general	Expedited hydropower license issuance under the condition that adaptive management be used at the site	Hydropower				·			FERC, 2000
Department of Agriculture	Monitoring	Implementation monitoring, baseline monitoring, validation monitoring, effectiveness monitoring	Natural resource conservation			■				USDA, 2003; USDA, 2003b;

Agency	Topic	Methods / Tools	Application Area	Elements of AM 1	2	3	4	5	6	Citation(s)
USDA Forest Service	Use of new decision-making techniques in adaptively managing the Bent Creek Experimental Forest in Asheville, NC	Multi-criteria decision analysis – analytical hierarchy process	Forest and terrestrial ecosystems		■	■	■	■		Rauscher *et al.*, 2000
California Coastal Commission	Performance evaluation for wetland mitigation projects	Passive adaptive management, well-defined monitoring program	Coastal and marine ecosystems		·	·				CCC, 1995
United States Navy	Application of adaptive management to the environmental cleanup of Navy facilities	Various analytical tools for evaluating remedy effectiveness and need for change	Remediation	■	■	■	■	■	■	NRC, 2003
United States Army Corps of Engineers	Use of ecological models and adaptive management in restoration of coastal ecosystems	System-development matrix	Coastal and marine ecosystems	·	■	·	■		·	Thom, 1997; Thom, 2000
	Decision-making framework for United States Army Corps of Engineers ecosystem restoration projects	Model-building; alternative restoration designs as "bet-hedging" strategy; multi-attribute decision analysis	General adaptive management		■	■	·	·		Pastorok *et al.*, 1997
	Guidance for monitoring within adaptive management framework for the United States Army Corps of Engineers	Framework for monitoring in adaptive management	General adaptive management			■	■	■		USACE, 2003
Fisheries and Oceans Canada	Socioeconomic framework for ecosystem-based fisheries management	Institutional analysis and development framework	Fisheries			·	■			Rudd, 2004
Health Canada	Adaptive management of emissions trading as part of mitigation measures for greenhouse gases	Stakeholder involvement and learning	Climate change management						■	CCHO, 2000
	Adaptive management in the guise of a health and air quality risk management framework	Stakeholder involvement at every step	Air quality risk assessment			·	·		·	Health Canada, 2003
Environment Canada	The ecosystem approach to natural resource management	Acknowledging uncertainty, learning about system	Biodiversity conservation					·		Environment Canada, 2004a

Agency	Topic	Methods / Tools	Application Area	Elements of AM						Citation(s)
				1	2	3	4	5	6	
	Environmental Assessment Best Practice Guide For Wildlife At Risk In Canada	Precautionary principle	Wildlife management			•	•	•		Environment Canada, 2004b
	Canadian National Wildlife Disease Strategy	Adaptive risk assessment and problem response framework	Wildlife management			■	■	■	■	Environment Canada, 2004c
British Columbia Forest Service	Framework and mechanics for adaptive management of forest ecosystems	Framework for running adaptive management workshops; solutions to common barriers	Forest and terrestrial ecosystems	■	•	■	■	■	■	Nyberg, 1999
	Examples of case studies in which the BCFS is applying adaptive management in the field	Management experiments to learn more about managed system	Forest and terrestrial ecosystems			■	■			BCFS, 2000
British Columbia Ministry of Forests	Reasons to implement adaptive management, problems with adaptive management, and tools for adaptive management of forests in British Columbia	Workshops, decision analysis, project design teams	Forest and terrestrial ecosystems	■		■	■	■	■	Taylor et al., 1997
British Columbia Ministry of Sustainable Resource Management	Land and resource management plan for the North Coast of British Columbia	Usage of up-to-date information and implementation of adaptive management	Forest and terrestrial ecosystems		■	•	■	■		BCMSRM, 2003
World Wildlife Federation	Problems with the European Union Common Fisheries Policy and suggestions for fixing them	Ecosystem-based management and the precautionary principle	Fisheries						■	WWF, 2001a
	Adaptive management and example applications in conservation	Adaptive management definition and framework	General Adaptive Management	■	■	■	■	■	■	WWF, 2001b

Six Adaptive Management Criteria: (1) Management objectives regularly revisited and accordingly revised; (2) A model(s) of the system being managed; (3) A range of management choices; (4) Monitoring and evaluation of outcomes; (5) A mechanism(s) for incorporating learning into future decisions; (6) A collaborative structure for stakeholder participation and learning

Meaning of Symbols: ■ = discussed in detail / demonstrated / performed in paper or case study reviewed by paper

• = mentioned or recommended but not explicitly carried out

Table 3.2. Adaptive management application papers.

Application Area	Topic	Methods / Tools	Funding Agencies	Elements of AM						Citation
				1	2	3	4	5	6	
Fisheries	Importance of learning in the risk management process with a case study involving water use for fisheries and hydroelectric power in Canada	Treating learning as an objective in collaborative stakeholder groups	University of British Columbia; Decision Research			·	·	■	■	McDaniels and Gregory, 2004
	Cooperation between small-scale users and management agencies in salmon management in British Columbia	Data sharing, building institutional capacity	Simon Fraser University					·	■	Pinkerton, 1999
	Prevention of shocks to developing fishery populations through adaptive management	Adequate monitoring of biological and economic fishery data	University of Washington; South Pacific Commission, New Caledonia		■	■	■			Hilbom and Sibert, 1988
	Simulation of adaptive management strategies for the reintroduction of trout cod	Stochastic population model, simulated monitoring uncertainty	Cooperative Research Center for Freshwater Ecology	·	■	■	·	·		Bearlin, 2002
	Simulating active vs. passive adaptive fisheries management in the Philippines	Ecological modeling software	Bedford Institute of Oceanography, Canada		■	■	■			Bundy, 2004
	The Decision Analysis and Adaptive Management (DAAM) Project and the Lake Erie walleye and yellow perch fisheries	Decision analysis to rank management alternatives and quantify uncertainty; adaptive management to reduce uncertainty	University of Guelph, Ontario		■	■	■	■	■	Nudds et al., 2003
Coastal and marine ecosystems	Wetland and finfish restoration	Restoration success index, passive adaptive management with thresholds for enacting pre-planned corrective measures	Public Service Electric and Gas Company		■	■	■			Weinstein et al., 1997
	Restoration of salt hay farm wetlands on the Delaware River Estuary	Evaluation of alternative channel construction techniques	American Society of Civil Engineers		■	·	·	·		Weishar and Teal, 1998

Satterstrom et al.

Application Area	Topic	Methods / Tools	Funding Agencies	Elements of AM 1	2	3	4	5	6	Citation
	Management of economic, environmental, social, and political aspects of large marine ecosystems	Five-module assessment and management methodology	Coastal and marine ecosystems				■		■	Duda and Sherman, 2002
Forest and terrestrial ecosystems	Modeling and selection of management plan for forest in west-central Alberta, Canada	Computer simulation of forest management scenarios	Alberta Health and Wellness; Millar Western Forest Products; Environment Canada; West Fraser Mills; Alberta Environment		■	■	■	·		van Damme et al., 2003
	Decision-making related to forest road deactivation in unstable terrain	Risk analysis	Forest Renewal British Columbia		■	■	■			Allison et al., 2004
	The need to consider pathogens, pests, and diseases in the adaptive management of forests	Computer modeling of pathogen distribution	National Research Foundation; Human Resource and Industrial Program; South African Forestry Industry	■						van Staden et al., 2004
	New system of environmental evaluation and its application in the Southern Appalachians	Novel environmental valuation system	National Science Foundation; Environmental Protection Agency		■	■	■	·	■	Norton and Steinemann, 2001
	Management of oaks and pines in the Wisconsin Necedah National Wildlife Refuge	Adaptive management framework	University of Wisconsin; Necedah National Wildlife Refuge	·	■	■	■	·		Haney and Power, 1996
	Evaluation of management of Costa Rican and Nicaraguan forests	Focus on the need for adaptive management	CATIE/FINIDA Research Fund; Center for International Forestry Research			·	·	·		McGinley and Finegan, 2003
	Ecosystem model-building for management of Southern bottomland hardwood forests	Delphi method of surveying experts	USDA Forest Service; Alabama Agricultural Experiment Station		■					Bliss et al., 1997
Florida Everglades	Adaptive management as a way to obtain the knowledge necessary for Everglades restoration	Ecological learning	University of Florida				·	■	■	Kiker et al., 2001

Application Area	Topic	Methods / Tools	Funding Agencies	Elements of AM						Citation
				1	2	3	4	5	6	
	Adaptive management in the Central and Southern Florida Project Comprehensive Review Study, a plan for restoration of the Florida Everglades	Review	University of California at Berkeley School of Law		■	■	■	■		Voss, 2000
	Use of risk assessment and adaptive management in the Everglades and South Florida ecosystems and the Central and Southern Florida Project Comprehensive Review Study	Risk-based conceptual model	University of Miami; United States Geological Society; South Florida Water Management District; Harwell Gentile & Associates		■	·	·			Gentile *et al.*, 2001
Rivers, other freshwater areas, and estuaries	Implementation of local watershed council management practices through adaptive management	Community-based conservation	Science to Achieve Results Graduate Fellowship, United States Environmental Protection Agency	·	·	■	·	·	■	Habron, 2003
	Assessment of information needed for environmental flow allocation, Lachlan River, New South Wales, Australia	Assessment of information needs for adaptive management	Macquarie University; Land and Water Resources Research and Development Corporation Project MQU6			■	■		■	Hillman and Brierly, 2002
	Gaps that hinder groundwater resource management in New Zealand	Case study	New Zealand Ministry of the Environment		■	■	■			Lowry *et al.*, 2003
	Application of the Plan for Analyzing and Testing Hypotheses (PATH) to the management of Snake River basin salmon stocks	Interagency collaboration, development of common data sets, reduction of uncertainty	Bonneville Power Administration		■	·	■	■	■	Marmorek and Peters, 2001

Application Area	Topic	Methods / Tools	Funding Agencies	Elements of AM 1	2	3	4	5	6	Citation
Wildlife management	Applying adaptive management to regulating waterfowl harvest	Passive and active adaptation algorithms	United States Fish and Wildlife Service; United States Geological Survey	■			■			Johnson and Williams, 1999
	Adaptive management of the species kokako to determine the effects of pests on population density	Variation of pest populations, monitoring of experimental and control areas, statistical analyses	Department of Conservation, New Zealand; Foundation for Research, Science, and Technology, New Zealand			■	■	·		Innes et al., 1999
	Conservation of biodiversity during natural gas exploration in Peru	Adaptive management with comprehensive monitoring	Smithsonian Institution; Shell International Limited	■		·	■	■	■	Dallmeier et al., 2002
	Integration of adaptive management strategies into habitat conservation plans	Incentives for adaptive management in HCPs	Washington Department of Natural Resources; Washington Department of Fish and Wildlife	■	■	■	■			Wilhere, 2002
	Framework for conservation of biodiversity using adaptive management to develop knowledge and skills	Various protection and management, law and policy, education and awareness, and incentive-changing strategies	Foundations of Success; Wildlife Conservation Society		·	·	■	■		Salafsky et al., 2002
Remediation	Modeling of rehabilitation of Cooum River, Chennai, India	Conceptual model building through workshops	Madras-Waterloo University Linkage Program; CUC-AIT	■	■	■	■		■	Bunch and Dudycha, 2004
	Restoration of damaged wetlands at San Diego Bay	Focus on mitigation compliance	San Diego State University			·	·	·		Zedler, 1997

Application Area	Topic	Methods / Tools	Funding Agencies	Elements of AM						Citation
				1	2	3	4	5	6	
General adaptive management	Models of optimization for harvesting policies for exploited populations	Theoretical adaptive optimization techniques	National Research Council, Canada; Environment Canada; International Institute for Applied Systems Analysis		■	·	■	■		Walters and Hilborn, 1978
	Gaps in the science base needed for ecological management, including gaps in adaptive management	Incentivizing cooperation	USDA Forest Service; Natural Resources Conservation Service; OMB; EPA; NOAA; US Army Corps of Engineers; US Department of Interior; Office of Science and Technology Policy	·		·			■	Szaro et al., 1998
	Principles of public involvement in adaptive management	Six propositions for integrating citizens into adaptive management	People and Natural Resources Program of the Pacific Northwest Research Station, USDA Forest Service						■	Shindler and Aldred Cheek, 1999
	Case studies of adaptive management related to land use, water quality, and recreational access	Software for policy option analysis	Macquarie University; Wyong Shire Council; University of Sydney		■	■	·		■	Gilmour et al., 1999
	Framework for adaptive ecological management and lessons learned from prior adaptive management attempts	Adaptive management framework from Nyberg, 1999	Society for Ecological Restoration International	·	■	■	■	■	■	Murray and Marmorek, 2003

Table 3.3. Adaptive management review papers.

Application Area	Topic	Methods / Tools	Funding Agencies	Elements of AM 1	2	3	4	5	6	Citation
Fisheries	Use of adaptive management to reduce uncertainty in fishery management and experiences with the Atlantic Canada groundfishery	Review of case studies	Natural Sciences and Engineering Research Council of Canada	·						Charles, 1998
	Discussion of gap between experimental marine ecology and fishery management; adaptive management as tool to bridge gap	Review	Presidential Chair in Science (Chile); Pew Charitable Fund Fellowship	·			·	·	·	Castilla, 2000
Coastal and marine ecosystems	Formal integration of adaptive management in coastal restoration through the Coastal Wetlands Planning, Protection, and Restoration Act (CWPPRA)	Reviews CWPPRA	Louisiana Department of Natural Resources			■	■	■	■	Steyer and Llewellyn, 2000
Forest and terrestrial ecosystems	Relationship between forests, agroforestry, biodiversity and using adaptive management to balance the three	Six key elements to biodiversity preservation in forests	The World Conservation Union						■	McNeely, 2004
	Decision support systems for federal forests in the United States	Decision support system software	USDA Forest Service		■	■		·		Rauscher, 1999
Florida Everglades	Uncertainty and the need for flexibility in adaptive Florida Everglades management	Stakeholder flexibility	John D. and Catherine T. MacArthur Foundation; Sustainable Everglades Initiative						■	Gunderson, 1999

Application Area	Topic	Methods / Tools	Funding Agencies	Elements of AM 1	2	3	4	5	6	Citation
Rivers, other freshwater areas, and estuaries	Gaps in knowledge about salinity management for freshwater biota	Review	Murray-Darling Basin Commission; National Action Plan Non-regional Foundation Funding		■					James et al., 2003
	Evaluation of estuary management through the National Estuary Program	National Estuary Program's Management Conference process	NOAA					■	■	Imperial et al., 1993
	Problems with implementing adaptive management in riparian ecosystems	Review / synthesis discusses challenges in each area	Natural Sciences and Engineering Research Council, Canada		■	■	■		■	Walters, 1997
	Adaptive co-management for freshwater resource management	Involving multiple groups as decision-makers with co-management	FORMAS, the Swedish Research Council for Environment, Agricultural Sciences and Spatial Planning					■	■	Folke, 2003
Wildlife management	Integration of wildlife management and research through adaptive management	Hypothetico-deductive science, learning as an objective	North Carolina State University; Colorado Division of Wildlife; National Biological Service; Utah State University; United States Forest Service; University of Guelph; Missouri Department of Conservation	■	■	■	■			Lancia et al., 1996
	Use of adaptive management in managing human-carnivore interaction	Behavior modification; isolation of humans from carnivores	Conservation International; Wildlife Conservation Society		■	■			■	Treves and Karanth, 2003
General adaptive management	Review of twenty-seven methodologies for assessment of management of protected areas	Collection of quantitative data from monitoring and qualitative data from scoring by managers and stakeholders	WWF/IUCN Forest Innovations Project				■			Hockings, 2003
	Framework for assessing the effectiveness of management of protected areas	"Measures of Success" framework developed by The Nature Conservancy and its partners	The Nature Conservancy				■			Parrish et al., 2003

4

SYSTEMS ANALYSIS AND ADAPTIVE LEARNING FOR PORTFOLIO MANAGEMENT OF SUPERFUND SITES

Peter A. Beling
University of Virginia Department of Systems Engineering

James H. Lambert
University of Virginia Department of Systems Engineering

Faheem A. Rahman
University of Virginia Department of Systems Engineering

George O. Overstreet
University of Virginia McIntire School of Commerce

David Slutzky
University of Virginia Department of Urban and Environmental Planning

ABSTRACT

This chapter develops a methodology to aid remedial project managers and senior management at the Environmental Protection Agency (EPA) in handling cleanup strategies and determining future reuse options at Superfund sites. The goal is achieved through the completion of four objectives: 1) avoiding back engineering and decisions that limit future reuse options; 2) understanding all contingencies in the cleanup process; 3) accounting for costs and benefits from cleanup and reuse for all stakeholders including future generations through the creation of an index of reusability; and 4) maximizing the opportunity for learning. Development of the method stems from concepts in risk and policy management previously applied in diverse disciplines. First, reuse contingency trees are developed to describe the potential deviations in the cleanup process that cause increased costs, increased time to reuse, or decreased future use options for a site. Second, a reusability influence diagram is developed to identify the impacts of the contingencies on the various actions and outcomes of the Superfund cleanup process. Third, an index of reusability is developed in order to compare the reuse benefits to cleanup costs and to herd portfolios of sites along different reuse strategies. The methodology is unique to any existing study of the Superfund process and the idea of the contingency binds the multiple methods. This chapter can lead to an initiative by the Environmental Protection Agency to move from the regional handling of sites to a broader national approach. Furthermore, the methodology has general importance not limited to Superfund sites but applicable to all environmentally impaired land.

INTRODUCTION

The Comprehensive Environmental Response, Compensation, and Liability Act (CERCLA), commonly referred to as the Superfund program, was passed as a way to respond to the impact of hazardous chemical wastes on human health. Under the program, the U.S. Environmental Protection Agency (EPA) is responsible for assessing the risks posed at the most hazardous waste sites and selecting the most appropriate cleanup strategy to protect human health and the environment, regardless of cost. In order to achieve this goal of protectiveness, the EPA has determined that the individual lifetime risk of cancer from exposure to contaminants should fall into a range of 1 in 10,000 to 1 in 1,000,000. Once this criterion is satisfied, the EPA considers the most cost-effective remedy where the Superfund program either funds the cleanup of the site, works with the state to clean up the site, or oversees cleanup by those responsible for the contamination.

Early in the Superfund program, the EPA was criticized for assuming too frequently that the future use of sites would be residential. Residential land use is the least restricted and allows for the greatest potential for exposure from activities performed on the land. This assumption typically led to two outcomes: 1) the sites were not cleaned up to the standards required, and/or 2) the cleanups at sites were extremely expensive. Cleaning up a site for residential use allows for the most reuse options in the future since additional cleanups may not be required. However, if the land is not used to its full unrestricted potential, it is an indication that the site was allocated too much funding, possibly taking resources away from cleaning other sites.

This chapter develops a methodology to aid remedial project managers (RPM), the EPA decision-makers, in handling cleanup strategies and determining future reuse at Superfund sites. The goal is achieved with the completion of four objectives: 1) avoiding dead-end decisions and back engineering; 2) understanding all contingencies in the cleanup process; 3) accounting for costs and benefits from cleanup and reuse for all stakeholders including future generations through the creation of an index of reusability; and 4) maximizing the opportunity for learning.

BACKGROUND OF SUPERFUND

From 1942 until 1953, the Hooker Chemical Company disposed of over 21,000 tons of chemical wastes in the Love Canal area of Niagara Falls, New York. Hooker covered the site with earth and clay and donated this land to the Niagara Falls Board of Education, which built schools and playgrounds on the site. In the summer of 1978, contamination began to surface on the schoolyard and seep into basements of local homes due to rising groundwater levels. The media brought attention to the problem and a public uproar ensued regarding the risks of hazardous waste sites across the nation.

On December 11, 1980, President Jimmy Carter signed into law the Comprehensive Environmental Response, Compensation, and Liability Act (CERCLA). CERCLA authorizes the Environmental Protection Agency (EPA) to remove or arrange for remedial action "whenever any hazardous substance is released or there is a substantial threat of such a release into the environment, or there is a release or substantial threat of release into the environment of any pollutant or contaminant which may present an imminent and substantial danger to the public health or welfare" (42 U.S.C. 9604). Removal is considered a short-term solution to control the release or threat of release of hazardous waste to "minimize or mitigate damage to the public health or welfare or to the environment" (42 U.S.C. 9601). By comparison, remedial action is considered as a long-term or "permanent remedy taken instead of or in addition to removal actions to prevent or minimize the release of hazardous substances so that they do not migrate to cause substantial danger to present or future public health or welfare or the environment" (42 U.S.C. 9601). CERCLA also provided for liability of parties responsible for releases of hazardous waste at sites. Additionally, a trust fund of $1.6 billion, commonly referred to as Superfund, was set up to pay for cleanup activity when responsible parties could not be identified. However, this money could only be allocated to sites placed on the National Priorities List (NPL), which contains the sites that pose the most significant potential threat to human health.

On October 17, 1986, Congress passed the Superfund Amendments and Reauthorization Act (SARA), which was a substantial amendment to CERCLA. This amendment had several key features: it increased the size of the Superfund to $8.5 billion; expanded the response authority of the EPA; strengthened enforcement activities at Superfund sites; broadened the law to include federal facilities; added a citizen suit provision; allowed the EPA to condemn property; and provided deadlines for response action. Also, SARA required the EPA to revise the Hazard Ranking System (HRS) to "ensure that the HRS accurately assesses the relative degree of risk to human health and the environment posed by hazardous waste sites and disposal facilities that may be placed on the NPL" (EPA 2003b).

SARA required the EPA to determine and consider the "applicable or relevant and appropriate requirements" (ARAR) from state and other federal environmental laws and regulations (42 U.S.C. 9621). It also emphasized the importance and preference of permanent cleanup and treatment at sites so as to reduce "volume, toxicity or mobility of the hazardous substances, pollutants, and contaminants" (42 U.S.C. 9621). This deterred the remedial action of on-site burial or transportation of contaminants to new sites where releases of the hazardous substances could continue. The statute also states that the EPA must take the following criteria into account when selecting a remedial action: "the long-term uncertainties associated with land disposal; the goals, objectives, and requirements of the Solid Waste Disposal Act; the persistence, toxicity, mobility, and propensity to bioaccumulate of such hazardous substances and their constituents; short- and long-term potential for adverse health effects from human exposure; long-term maintenance costs; the potential for future remedial action costs if the alternative remedial action in question were to fail; and the potential threat to human health and the environment associated with excavation, transportation, and redisposal, or containment" (42 U.S.C. 9621). After considering these, the EPA must select a remedy "that is protective of human health and the environment, that is cost effective, and that utilizes permanent solutions and alternative treatment technologies or resource recovery technologies to the maximum extent practicable" (42 U.S.C. 9621).

Superfund Process

Figure 4.1 is a layout of the process that a site undergoes from start to finish through the Superfund program. The figure is broken into several parts. The main phases of the process are located at the top of the figure: Pre-Characterization, Characterization, Remediation/Monitoring, and Use/Reuse. The next level of the figure, below the main phases, is the process flow. The squares indicate an event in the process and the arrows demonstrate the sequence of events. Events that take place for sites weeded out of the Superfund program are not included in this figure since inclusion of these events was outside the scope of the project. Finally, the bottom of the figure shows the average expected timeline for the process. According to 1989 calculations from a RAND study, from site awareness to construction completion (the EPA's term for the end of remedial action) takes an average of twelve years. Also based on that study, it takes eight and a half years to clean up a site once it has been placed on the NPL.

A site is first brought into the system from either public or state concern over contamination from air, soil, or groundwater. The EPA inventories the site within the CERCLA Information System (CERCLIS) database. Each site on CERCLIS undergoes a Preliminary Assessment (PA) and Site Inspection (SI). These initial screenings include examining existing documentation and some basic fieldwork to determine the nature and scope of the hazards. The EPA follows up with emergency removal actions at the site. From this event, a site can either be scored with the Hazard Ranking System (HRS) or designated that no further federal remedial action will be planned (NFRAP). States or cities take over the cleanup efforts at sites falling under the NFRAP designation. Sites that have scored sufficiently high on the HRS, meaning a score of 28.5 or higher, are placed on the NPL. HRS scores range from 0 to 100 and are "based on the likelihood that contaminants have been or will be released from the site, the physical and toxicological characteristics of the contaminants present at the site, and the human population or sensitive environments actually or potentially exposed to a release from the site" (U.S. Department of Energy 1994). Aside from the 28.5 cutoff for making sites eligible for the NPL, the EPA has not clarified the significance of the scores for taking remedial actions. While states or cities are responsible for handling sites not placed on the NPL, these sites can still "receive various Superfund actions without being on the NPL" (U.S. Congress, Office of Technology Assessment 1988).

Within the Characterization phase, the major process components are risk characterization, Reasonably Anticipated Future Land Use (RAFLU), Remedial Investigation and Feasibility Study (RI/FS), and Record of Decision (ROD). The RAFLU is the prediction on how the land at each site will be used. This event is important for gauging risk to future populations and for paring down cleanup alternatives to meet cleanup objectives. The RI/FS are two mechanisms for collecting data needed to determine the extent and nature of contamination at a Superfund site and for the development, screening, and detailed evaluation of alternative remedial actions. The ROD is a formal document that contains analysis of cleanup alternatives, expected costs, and the selection of the remedy to be used at a site.

These steps in the Superfund process are followed up by remedial design and action (RD/RA) in the Remediation and Monitoring phase. The RD is the phase where technical specifications for cleanup remedies and technologies are designed, while the RA is the actual implementation phase of the Superfund site cleanup. It is important to note that although there is no process specific to the Use/Reuse phase, it is still an extremely significant phase that must be recognized.

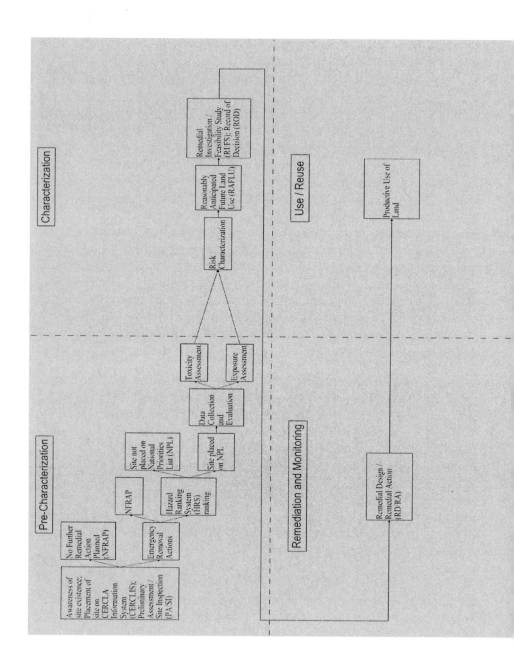

Figure 4.1. Superfund process flow.

Stakeholders in the Superfund Program

A stakeholder is any group or individual who can affect or is affected by the achievement of an organization's objectives. There are many stakeholders that affect and are affected, either positively or negatively, by the Superfund process. These stakeholders include: the local community, the site owner, the government or EPA, potentially responsible parties (PRPs), and present and future generations. Figure 4.2 (adapted from Dixon 1995) shows the major stakeholders and their interactions in the Superfund Process. The squares denote a stakeholder and a connecting line between two stakeholders indicates a direct relationship between the two.

The local community, consisting of residents on the site and/or close to the site and local government, interacts with the EPA, state governments, and participating PRPs to provide input on remedy selection, cleanup standards, and potential future land use. The EPA works with PRPs to negotiate cleanup funding and methodologies. The states set cleanup standards and provide input on the cleanup remedy selection. PRPs must determine liability among themselves. Passively participating PRPs merely write a check for the cleanup remedy. Both actively and passively participating PRPs recover costs from nonparticipating PRPs through litigation. PRPs seek compensation for cleanup and legal costs from insurers. Insurers seek compensation for their costs from reinsurers.

One of the biggest areas of academic debate regarding Superfund involves discussion of the liability structure. Liability for a Superfund remediation site is retroactive, strict, and joint and several. All present and past owners, transporters, and others can be liable for cleanup costs. Retroactive liability implies that a party can be held responsible even when their activities took place before CERCLA was passed. Strict liability means that these parties are accountable even when they were not negligent (Hird 1993). Finally, joint and several liability signifies that all parties are responsible for the cleanup costs in full or in part, depending on their ability to pay. For Superfund, there are unique incentives created by the liability structure, and there are unique constraints on the decision rules for treatment of the site.

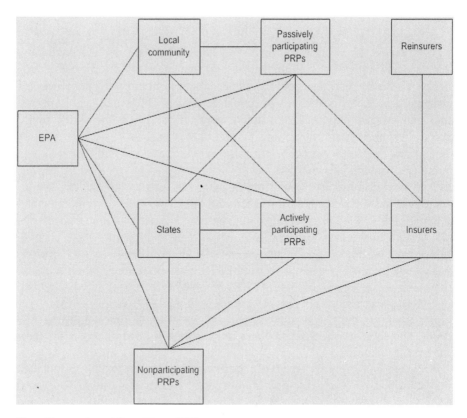

Note: Figure adapted from Dixon 1995

Figure 4.2. Stakeholders and interactions in the superfund process.

Reuse in the Superfund Program

There are currently over 1,500 sites on the NPL. NPL sites are subject to unique statutory procedures including requirements for certain remedial investigations, the provision of a formal record of decision, a period of implementation and monitoring, and a five-year review after construction is complete. "According to a 1994 analysis by the Congressional Budget Office, it takes on average twelve years to complete a Superfund cleanup (from the time of initial listing on the NPL to completion of a site's remedy)" (Wernstedt, Hersh & Probst 1998).

Thus far, 170 of the remediation sites in almost 40 states have been returned to productive use through Superfund. Over 60% of the sites have been redeveloped for commercial and industrial use with the remaining 40% of sites used for residential, ecological, recreational, agricultural, or governmental use. The following table (EPA 2003c) (Table 4.1) shows the exact breakdowns of land use, in terms of actual, planned, continued, and restored. According to the EPA (2003c),

> A site is in actual or planned use if a new commercial, residential, ecological, recreational, agricultural, governmental or other new use is occurring at the site, or if a detailed plan for a new use is in place. A site is in continued use if EPA has undertaken or has overseen the cleanup at the site, which allowed the site to be used productively during and after the cleanup. Restored use has occurred at a site when a preexisting use has been halted during cleanup, and was resumed after the site was cleaned up.

Table 4.1. Number of sites in productive use.

Note: Table from Environmental Protection Agency, Superfund Sites Returned to Productive Use Summary

Category	Primary Use						Totals
	Commercial	Residential	Ecological	Recreational	Agricultural	Governmental	
Actual Use	64	3	16	15	3	10	111
Planned Use	15	--	1	4	--	2	22
Continued Use	25	2	--	--	1	2	30
Restored Use	5	1	--	--	--	1	7
Totals	109	6	17	19	4	15	170

Under CERCLA, the EPA must "minimize or mitigate damage to the public health or welfare or to the environment" at Superfund sites (42 U.S.C. 9601). The EPA has three main objectives for these sites: 1) to protect human health from cancer risks, 2) to reduce non-cancer risks to individuals, and 3) to protect the environment. The EPA will operate within a range of acceptable risk when reviewing options for remediation. For example, remedial actions can only be taken at these sites if the action allows for only a 1 in 10,000 to a 1 in 1,000,000 individual lifetime risk of cancer from reasonable maximum exposure to contaminants after completion. Such a threshold becomes a goal of cleanup activities and also requires assumptions about future land use. The reuse of land at Superfund sites plays several roles in the remediation process. One role for reuse is as an input in the overall remedy selection. The EPA predicts and/or evaluates land use to estimate the likely risk at the site for negative human health or environmental effects. Another role for reuse is as a tool for risk management. From this point of view, land use is the outcome of remediation.

The following figure (Figure 4.3) demonstrates the role of land use assumptions as an input in the overall remedy selection. There are four main events in the Superfund process that are affected by the land use assumptions: remedial investigation, feasibility study, remedy selection, and remedial design/remedial action (Hersh, Probst, Wernstedt & Mazurek 1997). This figure is essentially a close-up view of the Superfund process flow figure (Figure 4.1), within the Characterization and Remediation and Monitoring phases. Land use for the remediation site is assumed at the first two events. At the remedial investigation event, land use assumptions allow for a baseline risk assessment. This assessment determines who will be at risk and by what pathways exposure may occur. An assumption of residential use may lead to higher values of carcinogenic risk. The same land use assumption is factored into the second stage, feasibility study, in order to determine remedial goals. These goals are numerical concentration limits for various site contaminants based on cancer and non-cancer risk.

In order to allow for less doctored risk assessments, seven factors are considered when evaluating the future land use at a site (Wernstedt, Hersh & Probst 1998). Figure 4.4 shows the seven factors involved in making this evaluation.

Over the long run, a site can have an interim reuse strategy. This gives the EPA time to determine a final reuse for the site while assigning short term uses that progressively allow for greater exposure to the land. There may be costs associated with using an interim strategy and changing to a final strategy. These costs may be classified as back engineering costs. For example, an interim use could be to use the land for recreational purposes. Later, the site may be changed to an industrial use site like a landfill. The costs associated with this change would be the high upfront costs of an expensive cleanup. This could also happen in reverse where a site is designated for industrial use but then the final use is determined to be recreational. In the second instance, the back engineering costs could be potentially higher costs for cleanup in the future.

Figure 4.3. Interaction of land use assumptions and the cleanup process.

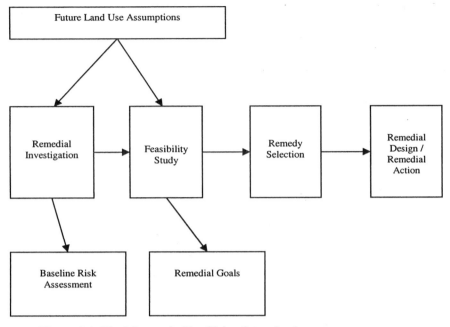

Figure 4.4. Ideal factors in identifying future land use.

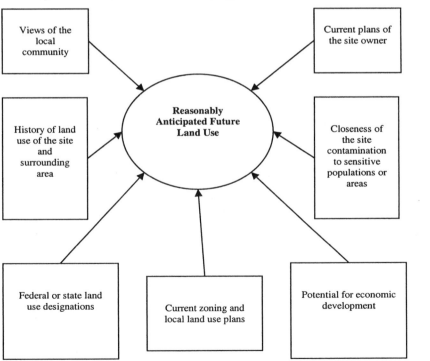

The Superfund Redevelopment Initiative (SRI), started in July 1999, "is EPA's coordinated national effort to facilitate the return of the country's most hazardous waste sites to productive use by implementing cleanup remedies that are consistent with the anticipated future use of the sites" (EPA 2003a). While EPA's main objective in the Superfund program is protecting human health and the environment, the EPA is taking the approach that the cleanups can be instrumental in returning contaminated sites to a usable, productive state. The reuse of the sites can range from restricted, where exposure of contaminants to populations is lower and the site can have a less extensive cleanup, to unrestricted, where the highest level of cleanup should be achieved to allow for maximum exposure to populations at or around the site. Restricted uses can include industrial or commercial uses, such as factories and shopping malls. Unrestricted uses can be for housing, public works facilities, transportation, and other community infrastructure. Additional unrestricted uses can be as recreational facilities, such as golf courses, parks and ball fields, or for ecological resources, such as wildlife preserves and wetlands (EPA 2003a). The EPA has increasingly recognized the need to determine what the future use of the site is likely to be, so that the EPA can achieve the secondary objectives of meeting cleanup standards for the anticipated future use and also apply cost-effective cleanup strategies.

THEMES OF THE PROJECT

In 2001, the University of Virginia opened the Center of Expertise in Superfund Site Recycling (CESSR) through a cooperative agreement with the EPA's Superfund Redevelopment Initiative (SRI). The overall mission of the multi-disciplinary CESSR is to "undertake investigations in support of the redevelopment of Superfund sites for a range of beneficial uses" (CESSR 2003). The goals of the CESSR are "to explore the relationship between site characterization and risk assessment, risk management, and private and public sector economics and decision-making in order to better predict the types of redevelopment that may be feasible and desirable under different conditions, and to facilitate that redevelopment" (CESSR 2003). The purpose and themes of this project align closely with the mission and goals of the CESSR.

The overall purpose for this project is to perform a systems analysis for particular reuse options in the management of the portfolio of Superfund remediation sites, by developing a decision and policy framework. The following four points are the themes that will be adhered to in our approach for managing environmentally impaired sites:

1) Avoid dead-end decisions and back engineering. A dead-end decision is one that limits or eliminates future options. Back

engineering is an attempt to recover previous options that are no longer available. There are substantial costs associated with back engineering. The basic idea is to preserve future options by making appropriate decisions in the present.

2) Understand all contingencies. Contingencies are events that cause deviations from the expected outcome of an action. In the context of the Superfund program, contingencies affect either the expected result of cleanup actions or future reuse options. Broad classifications of these contingencies include, but are not exclusive to: institutional uncertainties, physical/site parameters, human health, future reuse options, and technological availability. Understanding contingencies can allow remedial project managers (RPMs) to more effectively plan cleanup and reuse strategies at sites. Utilizing an adaptive management approach, RPMs can determine the effects of various contingencies and learn to mitigate the effects.

3) Account for costs and benefits from cleanup and reuse for all stakeholders including future generations. It is important to recognize future generations, since "the overwhelming preponderance of the risks is to future populations" (Hamilton & Viscusi 1995). The plan for this theme is to define and determine the as-planned outcome that the EPA attempts to achieve within the Superfund process. This as-planned outcome will provide an objective way to measure decisions resulting in alternate outcomes, allowing for the identification of dead-end decisions.

4) Maximize opportunities for learning by considering environmentally impaired sites at two levels. The first will be at the micro-level, where reuse options at individual sites are evaluated. At the macro-level, a portfolio of sites, such as Superfund remediation sites, is herded, applying principles of adaptive learning from past experience including successes and missteps. One way to herd these sites is by type, such as landfill, mining, and so on.

An important aspect of the effort is the development of an index of reusability that factors in the benefits and costs from reuse for Superfund sites. Although it may not be feasible to predict the likelihood of future use at a site, a numerical scale must be developed in order to compare sites according to reuse benefit potential and learn from the remediation process. Nothing in CERCLA or SARA suggests the maximization of net benefits or the achievement of optimal value, focusing instead on the protectiveness to human health and the environment. Also, neither directs cleanups to lead to the most productive use of Superfund sites. Instead, the statute emphasizes the minimization of costs within an accepted range of protectiveness (i.e., lifetime cancer risk).

We have considered how the remediation of environmentally impaired sites is presently carried out, and will now shift our focus to how it should be done. Our objective is to determine an as-planned outcome and the deviations associated with contingencies within a multi-criteria optimization framework. This is an enterprise where the interfaces of social and environmental systems must be understood and modeled in order to achieve well-advised outcomes. The difference between outcomes under an ongoing process and outcomes under other processes must be identified. The role of this project is to take a systemic approach to the Superfund program, which can be contrasted with the predominantly economic approaches that others have advanced.

ARTICULATION OF A SYSTEMS PERSPECTIVE

If the only objective of the Superfund program were to minimize cleanup cost, then the obvious solution would be to do nothing at zero cost. Clearly, this is not an optimal policy decision. A multiple objective framework is required in order to determine the optimal decision for Superfund site cleanups. "The concept of optimality in multiple objectives differs in a fundamental way from that of single objective optimization" (Haimes 2004).

Large-scale environmental problems or other public policy issues with many stakeholders rarely involve a single decision-maker. It is more common for multiple decision-makers to exist or for various stakeholders to influence the complex decision-making process (Haimes 2004). Each of the stakeholders has their own priorities and objectives for the decision being made. Within the Superfund program, the local community wants to eliminate the hazardous waste and the stigma associated with it from their neighborhoods. The potentially responsible parties (PRPs) want to minimize the cleanup cost and preserve or create the image of a corporation that protects the environment. Finally, the EPA wants to protect human health while maintaining cost-effectiveness of a cleanup and also returning the land to productive use. The focus of this project will be on the EPA's decision-making authority, accounting for its responsibility to manage the objectives of all other stakeholders.

There are four primary phases in the life cycle of site remediation: 1) pre-characterization, 2) characterization, 3) remedial action and monitoring, and 4) use/reuse. Each phase has three outcomes important to the decision-maker: costs, completion times, and performance, as depicted in the figure below (Figure 4.5). Each phase requires a specialized model of what factors drive the outcomes of the phase, and each phase is unique in the different classes of information that are available. Moreover, performance is a distinct concept for each of the phases. For example, the criteria for performance of risk characterization and sampling are different from the criteria for performance of a cleanup action. Furthermore, actions and outcomes in one phase affect the available decisions, uncertain inputs, and outcomes in subsequent phases. For most sites, however, there will be common driving factors of cost, completion time, and performance in each of the four phases.

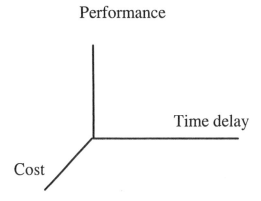

Figure 4.5. Three outcomes important for decision-making.

The current structure of the EPA, and therefore the Superfund program, divides the country into ten regions with each region handling the sites within its designated states. The flexibility of the program has its benefits. However, "case studies show the Superfund program as a loose assembly of disparate working parts; it is a system of divided responsibilities and dispersed operations. There is no assurance of consistently high quality studies, decisions, and field work or of active information transfer" (U.S. Congress 1988). In order to maintain the flexibility of the program, the Superfund program should continue making cleanup and reuse decisions within the ten region structure. However, the EPA must strive for a broader level of observation and data collection in order to allow for the utilization of adaptive management techniques. Rapid data collection and distribution from a national perspective can ensure consistency in cleanups among regions, ensure the efficient use of funds, and provide opportunities for learning from successes and missteps from one site to another.

A broader view of the Superfund program, combined with an emphasis on the productive use of the sites, allows the focus of the driving factors to fall on the last phase of the process: use/reuse. In this phase, performance can be equated with benefits in the use/reuse phase, with cost treated as very simply the cost associated with the use/reuse phase. Finally, the time delay component can actually be removed as an explicit driving factor and considered implicitly within the benefits and costs. The EPA's goals with respect to the driving factors are to maximize benefits and minimize costs. Since this phase occurs after the remediation/monitoring phase, the assumption has been made that the desired level of protectiveness to human health and the environment has been achieved.

CONTINGENCIES

This section begins to detail several of the methodological concepts applied in this chapter. The following figure (Figure 4.6) provides a graphical overview of the concepts. First, reuse contingency trees (Part A of Figure 4.6) are developed to describe potential deviations in the cleanup process that cause increased costs, decreased benefits, or decreased future use options for a site. Second, a reusability influence diagram (Part B of Figure 4.6) is developed to identify the impacts of contingencies on the various actions and outcomes of the Superfund cleanup process. Third, an index of reusability (Part C of Figure 4.6) is developed in order to compare reuse benefits to costs and to herd portfolios of sites along different reuse strategies. This section will focus on the reuse contingency tables and trees and the reusability influence diagram. The next section continues the application of concepts with a discussion of the benefits and costs associated with reuse and cleanup and the creation of a reusability index.

A key objective for this project is to identify and understand all contingencies associated with sites that undergo remediation through the Superfund process. Contingencies are events which cause deviations from the expected outcome of an action. In the context of the Superfund program, contingencies affect either the expected result of cleanup actions or the future reuse options. Broad classifications for these contingencies include, but are not exclusive to: institutional uncertainties, physical/site parameters, human health, future reuse options, technological availability, and learning potential.

Table 4.2 was adapted from a similar methodology applied to the Springfield Interchange Improvement Project by the Virginia Department of Transportation. That project has a timeline of twelve years and has an estimated cost of approximately $700 million. In the financial plan for the Springfield Interchange project, there is a table that summarizes key project assumptions, risks, and risk mitigation. The following table provides a similar approach and applies the concept to the Superfund project. These contingencies were determined by reviewing approximately twenty case studies of Superfund sites and by communicating with experts familiar with

A. Reuse Contingency Tree

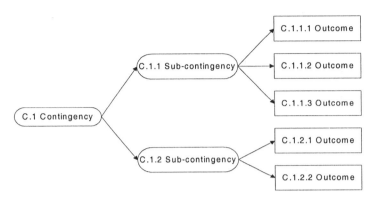

B. Reusability Influence Diagram

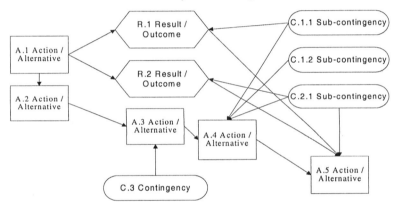

C. Reuse Benefit - Cost Analysis

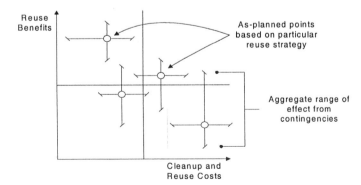

Figure 4.6. Overview of methodology.

the process. Each of the existing case studies was written from the perspective of what the author of the case was studying, including incorporation of future use into the cleanup process, the use of institutional controls, and local community involvement. Compiling this list was a major challenge since a system-wide view of the potential deviations has never before been undertaken.

Table 4.2. Sources of deviations and contingencies.

Contingency ID	Contingency Outcome	Contingency Classification	Primary Stakeholders	Outcome Mitigation	Notes
C.1.1.3	Cleanup efficacy is worse than expected	Technology - Efficacy	EPA, States, Local community	Learn effectiveness from application of technology on other sites	---
C.1.2.1	Cleanup cost is greater than expected	Technology - Cost	EPA, PRP	Learn cost accuracy from application of technology on other sites	---
C.1.3.2	PRP disagrees with selected cleanup remedy	Technology - Stakeholder Participation	EPA, PRP	N/A	PRP is motivated to minimize cost of cleanup
C.1.3.3	Local community disagrees with selected cleanup remedy	Technology - Stakeholder Participation	EPA, Local community	N/A	Local Community is motivated by the permanence of the cleanup
C.2.1.1	Shift in market and demand for land use	Future Reuse - Market	EPA, States, Local community	N/A	Anticipated future land use no longer reasonable
C.2.2.2	PRP does not agree with anticipated use	Future Reuse - Stakeholder Participation	EPA, PRP, Local Community	Allow PRP to lead independent study of contamination and anticipated land use	---
C.2.2.3	Local community does not agree with anticipated use	Future Reuse - Stakeholder Participation	EPA, PRP, Local Community	Open channel for communication of united public opinion	---

Table 4.2. Sources of deviations and contingencies (cont.).

Contingency ID	Contingency Outcome	Contingency Classification	Primary Stakeholders	Outcome Mitigation	Notes
C.3.1.1	Contaminants more harmful than originally percieved	Human Health - Cancer Risk	EPA, Local community	N/A	---
C.4.1.1	Greater concentration of contaminants on site than originally determined	Physical / Site Paramter - Contamination	EPA, Local community	N/A	---
C.5.1	Site is highly contaminated but does not achieve HRS score threshold	Institutional - NPL Listing	EPA, States	N/A	---
C.5.2.2	Institutional control and remedy preclude future use	Institutional - Controls	EPA, Local community	N/A	Example: Capping and zoning a site for industrial use and then realizing nothing can be built on cap
---**	Modest cleanup effort push additional costs out into future for a more extensive cleanup	Technology - Cost	EPA, PRP	N/A	---
---**	Site no longer qualifies for federal remedial action	Institutional	EPA, States	Site responsibility turned over to state	---

** Since these have no Contingency ID, they are not displayed on the Reuse Contingency Trees

The headers in the previous tables include: Contingency ID, Contingency Outcome, Contingency Classification, Primary Stakeholders, Outcome Mitigation, and Notes. The contingency ID corresponds to where the entry can be found on the reuse contingency tree (Figure 4.7). The reuse contingency tree is described in detail later in this section. The contingency outcome is an event that occurs at some point during the Superfund process. Contingency classification is the broad category under which the specific contingency falls. The categories include, but are not exclusive to: Technology, Human Health, Physical/Site Parameter, Institutional, and Future Reuse. From the previously completed stakeholder analysis (Figure 4.2), it is known that all stakeholders have an effect on each other. This means that each contingency will have an effect on every stakeholder. Therefore, the entry in the table for primary stakeholder indicates the stakeholders which benefit or lose as a direct result of the contingency occurring. For example, a cost overrun in cleanup directly affects the EPA and PRPs. However, the local community at another site can be affected since less budget money will be available for a cleanup near their site. The other stakeholders will also be affected, but since these are indirect consequences of the contingency, they are not included in this column. Since these contingencies were compiled from the perspective of the EPA, every contingency has the EPA listed as a direct stakeholder. However, the table is generic enough to allow for contingencies to be compiled from different stakeholder perspectives. The next two columns were inserted to provide guidance and clarification for remedial project managers (RPMs) and others using the contingency table. Outcome mitigation provides methods to reduce the effect of the contingency on the expected outcome of the cleanup and reuse process. These were compiled based on the experiences of RPMs found in the case studies. Lessons learned about contingencies at sites across the country now have a central location where RPMs can relate their experiences to one another. Entries with "N/A" indicate that either there is no specific way to mitigate the outcome or that none was found in the case studies. Finally, the notes column is where greater detail and clarity can be found for each of the contingencies.

The next step was to create reuse contingency trees in order to capture every possible outcome that could occur from a potential deviation. Figure 4.7 is the result of this creation. In the figure, the ovals represent contingency categories and contingency subcategories. The rectangles symbolize the possible outcomes that result from the contingencies. These contingencies were determined by reviewing approximately twenty case studies on Superfund sites and communicating with experts familiar with the process.

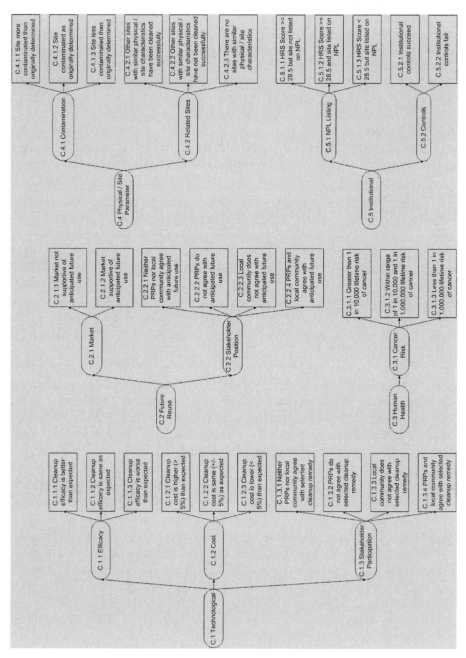

Figure 4.7. Reuse contingency trees.

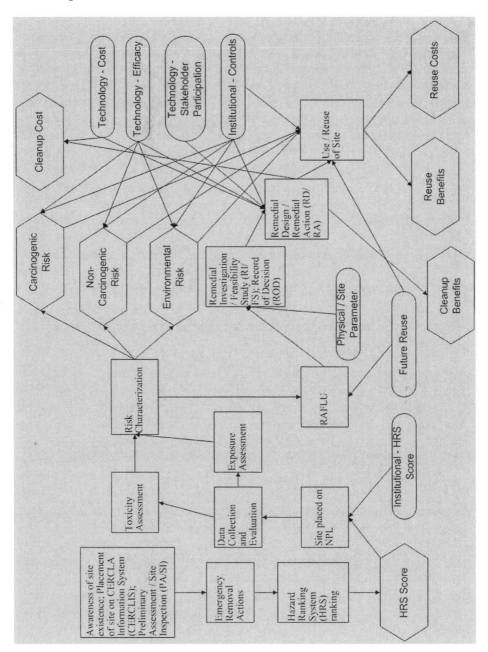

Figure 4.8. Reusability influence diagram.

Figure 4.8 illustrates the use of an influence diagram for site remediation decisions. An influence diagram is a compact, graphical model of a decision. The influence diagram has an underlying quantitative basis in relationships among the set of alternatives, the uncertain or random states of nature, and the outcomes for the decision-maker. The figure links the contingencies to the major phases of cleanup: pre-characterization, characterization, remediation and monitoring, and use/reuse. Notice in the figure the representation of the available options or actions (rectangles), uncertain inputs, referred to here as contingencies or potential deviations (ovals), and uncertain outcomes (hexagons) within a particular phase of the remediation effort. Rectangles are decision nodes, which represent a set of choices, actions, or alternatives. Arrows into decision nodes represent information that is available at the time of the decision. Ovals are uncertain inputs, which represent a probability distribution over the potential outcomes listed in the contingency trees. In the context of the Superfund program, the uncertain inputs are synonymous with the contingencies that occur throughout the process. Hexagons are value nodes, which represent a measure of the outcome of the decisions and states of nature. To acknowledge multiple objectives, such as benefits and costs, corresponds to including more than one value node in the diagram.

COST-BENEFIT ANALYSIS

Cost-benefit analysis was used to evaluate options for fixed roadway lighting. In the lighting problem, the benefits of adding lights were compared to the costs with respect to accident rates and traffic volume (Lambert, Baker & Peterson 2001). A similar analysis can be conducted for Superfund sites in terms of site reusability. The benefits of reusability can be compared to the costs for cleanup and productive use of the site.

The cost for cleanup and reuse can be broken down into several categories, as seen in Figure 4.9: cleanup technology, operation and maintenance, implementation of reuse and institutional control regulation. The first two categories, cleanup technology and operation and maintenance, are functions of the cleanup technology selected and the concentration of each contaminant. Institutional control regulation is a function of which institutional controls are selected and the desired land use. Typically, the EPA and PRPs handle the cleanup technology and the operation and maintenance costs. Occasionally, the states are responsible for a portion of the operation and maintenance costs. Finally, the states and local communities are accountable for the placement and upkeep of institutional controls. One important cost unaccounted for in the figure is the long-run cost for the change in productive use.

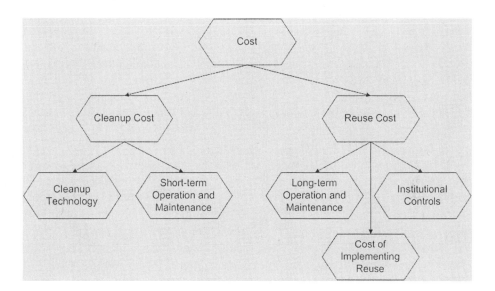

Figure 4.9. Cleanup and reuse costs.

Recognizing the benefits of site remediation presents a greater challenge since many of the benefits are realized at varying degrees and different times well after cleanup has been completed. According to an unpublished study completed by E^2 Inc., there are five major areas of benefits: human health cost savings, property value enhancement due to reuse, agricultural reuse, recreational reuse, and smart growth from reuse (E^2 Inc. 2001). Cost savings for each health case avoided are derived as a direct result of cleanup activity, whereas the other benefits are directly obtained from remediation and reuse of the land. Therefore, the benefits can be broken into two broad categories: benefits before reuse and reuse benefits. The following figure (Figure 4.10) displays the categories and areas for the benefits before and after reuse, both of which are results of the cleanup. The figure displays additional areas of benefits; however, the benefits shaded in gray are those that will be focused on throughout the remainder of the chapter.

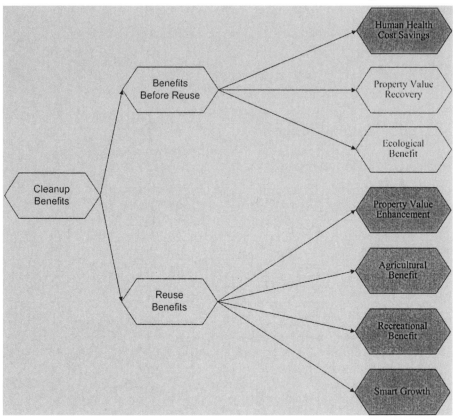

Figure 4.10. Benefits from cleanup and reuse.

Human health cost savings can be broken down into more specific categories: injuries and deaths avoided from acute events, birth defects avoided, cancer cases avoided, mental impairment from lead exposure avoided, and other miscellaneous cases avoided. Table 4.3 provides cost savings equations for each of these categories. These equations are adopted from the E^2 Inc. study. Although individual health benefits have been studied previously, this report "provides the first comprehensive approach to the assessment of multiple health benefits."

Benefits from human health cost savings are calculated by multiplying the population within a one-mile radius of the site by the number of cancer cases, birth defects, mental impairments, and miscellaneous health cases and adding

Table 4.3. Human health cost savings.

Category	Cost Savings Equation	Resulting Units
Injuries and deaths avoided (ID)	ID = (DA)(VL) + (IA)(CI) where DA = number of deaths avoided VL = statistical value of life IA = number of acute injuries avoided CI = average medical cost of treating acute injury	$
Cancer cases avoided (CA)	CA = (EC)(CT) where EC = excess cases of cancer avoided CT = average cost of cancer treatment	$ / person
Birth defects prevented (BD)	BD = (BR)(EB)(CD) where BR = birth rate around site EB = excess birth defects avoided CD = average cost of handling birth defect	$ / person
Mental impairment cases avoided (MN)	MN = (PC)(MC + ED) where PC = percentage of children around site MC = medical cost of mental impairment ED = educational cost of mental impairment	$ / person
Miscellaneous cases prevented (MS)	MS = (MI)(CM) where MI = number of miscellaneous health cases CM = average cost of treating miscellaneous case	$ / person

Note: Equations from E^2 Inc. 2001

the result to the amount saved from injuries and deaths avoided. The following equation summarizes the cleanup benefits:

$$CB = [(EP)(CA + BD + MN + MS)] + ID$$

where CB = cleanup benefit before reuse
EP = exposed population within a one-mile radius of the site
CA = cancer cases avoided
BD = birth defects prevented
MN = mental impairment cases avoided
MS = miscellaneous cases prevented
ID = injuries and deaths avoided

The reuse benefits derived from site remediation are more straightforward since they are less specific. This can be attributed to the fact that less is known about these benefits. Property value benefits are calculated by determining the increase in land value both on and off site in the surrounding areas. Agricultural reuse benefits are given based on the value of services the land provides. Recreational reuse benefits are given based on the potential revenue generated. Finally, smart growth reuse benefits are defined by the E^2 Inc. study as "the market-priced benefits provided by open space protection measures and infill development." Table 4.4 summarizes the benefit equations for each of these areas. The following equations are also adopted from the E^2 Inc. study.

Table 4.4. Site remediation benefits.

Catergory	Reuse Benefit Equation	Units
Property Value Enhancement (PR)	PR = (MN)(MP) + (PA - PB) where MN = marketable land on site MP = market price PA = price of offsite land after reuse PB = price of offsite land before reuse	$
Agricultural Reuse (AR)	AR = (AE)(SV) where AE = acres of land for agricultural reuse SV = service value	$
Recreational Reuse (RR)	RR = (AR)(PN) where AR = acres of land for recreational use PN = potential revenue	$
Smart Growth Benefit (SG)	SG = (AG)(VN) where AG = acres of greenfield development offset VN = value of land	$

Note: Equations from E^2 Inc. 2001

The reuse benefit is obtained by summing the different components. The following equation demonstrates the reuse benefits:

$$RB = PR + AR + RR + SG$$
where RB = reuse benefit
PR = property value enhancement
AR = agricultural reuse
RR = recreational reuse
SG = smart growth benefit

The overall benefit is equal to the sum of the cleanup benefits, which in the scope of this chapter are the human health cost savings, and the reuse benefits. The E^2 Inc. study calculated the expected overall benefit of the Superfund program represented as an annualized net present value. This project seeks to extend the utility of calculating the benefits and costs for each Superfund site that is part of an overall cleanup and reuse strategy. First, the cleanup benefits (CB) will be considered a constraint to achieving productive use. The assumption is being made that a certain level of protectiveness must be achieved at every site. Therefore, a minimum level of cleanup benefits is expected at each site. If this level is not accomplished, then productive use at the site cannot begin.

It is not useful to look at the specific dollar values produced by the reuse benefit and cost equations, due to the uncertainty in determining exact values for the elements comprising the benefits and costs. Therefore, an index of reuse benefits and an index of costs were developed. The following equation demonstrates the cost index:

$$TCI_j = \frac{TC_j}{(\frac{1}{n})\sum_{i=1}^{n} TC_i}$$

where TCI_j = total cost index for site j

TC_j = expected total cost at site j for a particular reuse

$(\frac{1}{n})\sum_{i=1}^{n} TC_i$ = average total cost across a portfolio of sites for the particular

reuse

$$j \notin (1,2,3,...,n)$$

The cost index is the expected total cost for a particular reuse at a given site divided by the average cost of that same reuse across a portfolio of sites that have already been cleaned and returned to productive use. The average is flexible in the sense that the portfolio can be identified in any way the user wants. For example, the average can be the national average across all sites with that particular reuse, across sites with similar characteristics, or across sites within the same region.

The reuse benefit index is calculated in a similar fashion by using the following equation:

$$RBI_j = \frac{RB_j}{(\frac{1}{n})\sum\limits_{i=1}^{n} RB_i}$$

where RBI_j = reuse benefit index for site j

RB_j = expected reuse benefit at site j for a particular reuse

$(\frac{1}{n})\sum\limits_{i=1}^{n} RB_i$ = average reuse benefit across a portfolio of sites for the

particular reuse

$j \notin (1,2,3,...,n)$

The reuse benefit index is the expected reuse benefit for a particular reuse at a given site divided by the average of reuse benefits of that same reuse across a portfolio of sites that have been remediated and returned to productive use. Once again, the average is flexible. However, it is important to realize that the same sample for averaging must be used in both equations in order to accurately compare the two indices. Since the Superfund program is still young, there will be only a small sample of sites to use initially for the average. This average will stabilize over time after more sites have been returned to productive use and the costs and reuse benefits have been realized.

Finally, the cost index and the reuse benefit index can be combined to form the reusability index, which is simply the ratio of the reuse benefit index to the cost index. The following equation summarizes the idea:

$$\Theta_j = \frac{RBI_j}{TCI_j}$$

where Θ_j = reusability index for site j

The next step is to understand how to use these indices. Figure 4.11 provides a demonstration. In the graph, the reuse benefit index and the cost index are each on an axis. Each site will have a set of cost indices and reuse benefit indices for each available reuse option. The preferred options are those which have high benefits and low costs. The graph can also be used for a portfolio of sites in a similar fashion. Separate reuse strategies for a portfolio, each consisting of one reuse option for each site in the portfolio, can be plotted on the graph and compared to one another to find the preferred reuse strategy.

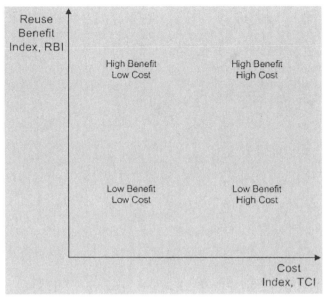

Figure 4.11. Reusability index.

APPLICATION OF CONCEPTS

In order to understand the role of contingencies in this cost-benefit analysis, an example is presented. Assume that there are four Superfund sites W, X, Y, and Z. Site W is a 100-acre mining site with contaminants including lead, zinc, and copper tailings. The surrounding land is used for commercial and light industrial purposes. Site X is a 10-acre landfill site contaminated with volatile organic compounds. Site Y is a 50-acre mining site, similar to site W. Finally, site Z is a 500-acre former military facility.

RPMs and senior managers must determine the reuse strategies for these sites individually and as a whole, due to budgetary constraints. Figure 4.12 demonstrates the methodology can be used for a single site, site W, by graphing the reuse benefit indices and cost indices associated with various reuse options and identifying the affects of contingencies. From this graph, RPMs can see that residential use for this site has low benefits and high costs with respect to the average benefits and costs normally associated with residential use. The other three choices, recreational, commercial, and industrial, have a reusability index of at least 1, indicating that every dollar spent on cleanup and reuse is rewarded with at least a dollar in reuse benefits. The effects of the contingencies on each of the reuse options are displayed as the horizontal and vertical lines protruding from the option as well as the shaded region around the reusability index points for each reuse option. Although in this example, industrial use may seem like the optimal choice for site W, it is important to realize that this methodology does not seek to optimize the reuse options for individual sites. The focus of the methodology is to provide objective assistance to the decision-makers and not to make the decision.

Reuse	Reuse Benefit Index			Cost Index			Reusability Index
	Low	Middle	High	Low	Middle	High	
Industrial	1.35	1.60	1.80	0.20	0.60	0.80	2.67
Commercial	0.65	1.20	1.35	0.90	1.20	1.50	1.00
Recreational	0.45	0.80	1.25	0.50	0.75	1.05	1.07
Residential	0.15	0.45	1.15	1.25	1.70	2.00	0.26

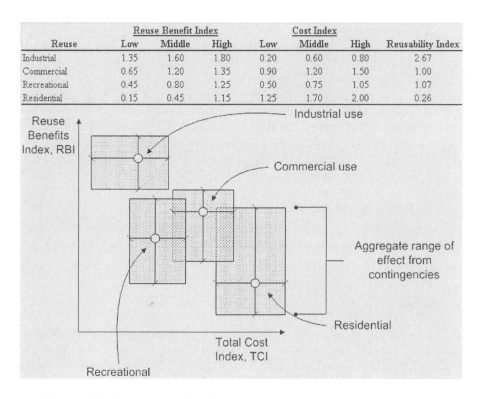

Figure 4.12. Reuse options for Site W.

In Figure 4.13, RPMs and senior managers can view multiple reuse strategies across the portfolio of sites W, X, Y, and Z. By using different combinations of reuse options with high reusability indices for individual sites, multiple reuse strategies can be developed. The figure demonstrates one of these reuse strategies. The layout of the graph and the data table are exactly the same as for plotting the single site reuse options. Notice that the reusability index for the industrial use of site W corresponds to the same point on Figure 4.12. The basic idea behind each of the reuse strategies is that decision-makers can provide constraints, such as annual budgetary limitations, in order to limit the number of combinations of reuse strategies for a portfolio of sites.

Reuse	Reuse Benefit Index			Cost Index			Reusability Index
	Low	Middle	High	Low	Middle	High	
Industrial for site W	1.35	1.60	1.80	0.20	0.60	0.80	2.67
Recreational for site X	0.10	0.35	0.80	0.15	0.60	1.10	0.58
Commercial for site Y	0.80	1.15	1.35	0.45	0.70	0.90	1.64
Commercial for site Z	0.80	1.05	1.50	1.00	1.50	1.95	0.70

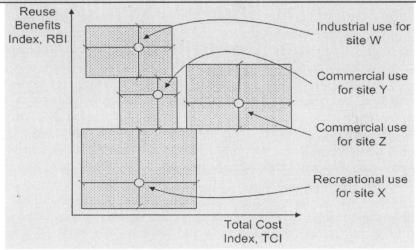

Figure 4.13. Sample reuse strategy for Sites W, X, Y, and Z.

CONCLUSION

Cost-benefit analysis is a framework to provide structure when evaluating decisions, a framework that requires attention to determine the impacts of contingencies on decisions and assigning values to the outcomes (Stokey & Zeckhauser 1978). As a tool, it provides only a means for opening an objective decision-making procedure, not for making the actual decisions. The usefulness of the model lies not in projecting actual costs or benefits of the Superfund program, but in understanding the dynamics of the system with respect to the contingencies. Understanding the dynamics of the system can help EPA decision-makers to anticipate, understand, and deal with most problematic contingencies of the Superfund program.

The contingency-based modeling effort is beneficial to several stakeholder groups within the EPA. The model is useful to RPMs because it clarifies all potential contingencies and allows them to learn from each other's experiences. Also, the RPMs and senior management at the EPA benefit by avoiding the political nature in which sites are prioritized for cleanup. The model can facilitate potential benefits impact analysis and allow for an objective ranking of sites where the sites with the highest potential reuse benefit can have cleanup strategies applied first.

The EPA can also use the model to motivate other stakeholders to invest in cleanups. This is done by tying the contingencies, and the stakeholders either responsible for them or affected by them, to the reuse benefits at Superfund sites. PRPs prefer to delay cleanups since a dollar spent tomorrow is cheaper than the same dollar spent today due to present value discounting. However, the EPA can use the model to prevent the time delay contingency from occurring by demonstrating to the PRP that the reuse benefit for that site is also affected by delaying the cleanup. In these cases, the PRPs are recipients of the reuse benefits but they cannot realize these benefits until completion of the cleanup. This varies from the current approach of CERCLA, which requires only that the sites are protective of human health and the environment. Using this model, the EPA would use a combination of the protectiveness and reuse benefit arguments to prevent the PRPs from delaying the cleanup at a site.

Institutional control failure is a relatively new contingency that has become important for ensuring particular reuse options at Superfund sites. The EPA can use the model to motivate the states that fund institutional controls at a site to continue to maintain the controls by demonstrating the potential loss of reuse benefit if the controls are not maintained.

This chapter is just the first step in developing the methodology of promoting productive use not just at individual sites, but across the Superfund program as a whole. The developed methodology makes several assumptions about the long-term state of the Superfund program. First, a more thorough understanding of the reuse benefits and costs needs to be achieved. Second, data have to be gathered from many sites that have completed cleanup and are undergoing reuse to provide a good sample for determining average reuse benefits and costs. Table 4.5 provides a list of required data elements based on the cost and benefit equations found in the chapter. Many of the data elements will have to be approximated since their location is not yet determined or simply cannot be found. However, the data will be useful in predicting the expected reuse benefits and costs for sites still to be remediated and returned to productive use. Third, the effect of contingencies during the cleanup process on reuse benefits should be fully explored. Once a range of data on reuse benefits and costs has been collected, an impact analysis of the contingencies should be performed. After satisfying the assumptions, the developed methodology must be tested, verified, and tuned. Finally, the best ways to herd a portfolio of sites should also be tested.

In the long run, for this methodology and the Superfund program as a whole, it would be interesting to observe how multiple future uses for a single site can be modeled. This means that the first reuse can be as an industrial landfill, but maybe fifty to a hundred years later the most beneficial use of the land could be residential. A follow-up question would be: Are the PRPs or the Superfund program responsible for funding the change in productive use? Currently, there is no research on this problem, probably because the Superfund program is still in its early phases.

Table 4.5. Data elements required for methodology.

Data Category	Data Element	Source	Notes
Cleanup cost	Type and level of each contaminant on site	CERCLIS	---
Cleanup cost	Cleanup cost	TBD	Function of technology and type and level of contaminant
Cleanup benefit before reuse - Human health cost savings	Exposed Population	E^2 Inc.	E^2 Inc. is compiling this data for different years and varying radii around each Superfund site
Cleanup benefit before reuse - Human health cost savings	Statistical value of life	E^2 Inc.	E^2 Inc. is compiling this data for different years
Cleanup benefit before reuse - Human health cost savings	Number of lives saved (or deaths avoided)	TBD	Estimate based on sites with similar characteristics
Cleanup benefit before reuse - Human health cost savings	Number of injuries prevented	TBD	Estimate based on sites with similar characteristics
Cleanup benefit before reuse - Human health cost savings	Average savings per injury prevented	TBD	---
Cleanup benefit before reuse - Human health cost savings	Excess cancers avoided	TBD	Estimate based on sites with similar characteristics

Table 4.5. Data elements required for methodology (cont.).

Data Category	Data Element	Source	Notes
Cleanup benefit before reuse - Human health cost savings	Average cost per cancer treatment saved	TBD	---
Cleanup benefit before reuse - Human health cost savings	Birth rate	TBD	---
Cleanup benefit before reuse - Human health cost savings	Excess birth defects avoided	TBD	Estimate based on sites with similar characteristics
Cleanup benefit before reuse - Human health cost savings	Average saving per birth defect avoided	TBD	---
Cleanup benefit before reuse - Human health cost savings	Percentage of children in population (for each site)	Risk and exposure assessment	---
Cleanup benefit before reuse - Human health cost savings	Medical and educational savings from mental impairment cases avoided	TBD	---
Cleanup benefit before reuse - Human health cost savings	Excess cases of miscellaneous health problems	TBD	Estimate based on sites with similar characteristics
Cleanup benefit before reuse - Human health cost savings	Average savings per miscellaneous health problem	TBD	---
Reuse Benefit - Property value enhancement	Marketable land on site	TBD	---

Table 4.5. Data elements required for methodology (cont.).

Data Category	Data Element	Source	Notes
Reuse Benefit - Property value enhancement	Market price of land on site	Tax records	Lack of consistency for property valuation
Reuse Benefit - Property value enhancement	Price of off site land before reuse	Tax records	Lack of consistency for property valuation
Reuse Benefit - Property value enhancement	Price of off site land after reuse	Tax records	Lack of consistency for property valuation
Reuse Benefit - Ecological Reuse	Acres of land for ecological use	RAFLU	Estimate based on sites with similar characteristics
Reuse Benefit - Ecological Reuse	Service value per acre	TBD	Estimate based on sites with productive use
Reuse Benefit - Recreational Reuse	Acres of land for recreational use	RAFLU	Estimate based on sites with similar characteristics
Reuse Benefit - Recreational Reuse	Potential revenue	TBD	Estimate based on sites with productive use
Reuse Benefit - Smart Growth	Acres of greenfield development offset	TBD	---
Reuse Benefit - Smart Growth	Value of land	E^2 Inc.	---

REFERENCES

Brazell, D., and G. Gerardi. 1994. "Issues in Financing the Superfund." *National Tax Journal* 47 (3): 677-688.

Breggin, L., J. Pendergrass, and J. van Berg. 1999. *Protecting Public Health at Superfund Sites: Can Institutional Controls Meet the Challenge?* Washington, DC: Environmental Law Institute.

Buede, D. M. 2000. *The Engineering Design of Systems: Models and Methods.* New York: John Wiley & Sons, Inc.

CESSR. 2003. Center of Expertise in Superfund Site Recycling. http://www.virginia.edu/ superfund Accessed 1 July 2003.

Dixon, L. 1995. "The Transaction Costs Generated by Superfund's Liability Approach." In Revesz and Stewart 1995a: 171-186.

E^2 Inc. 2002 (August). Superfund Benefits Analysis DRAFT. Unpublished study for EPA, Office of Emergency and Remedial Response.

Environmental Law Institute. 1995. *Institutional Controls in Use.* Washington, DC: Environmental Law Institute.

Environmental Protection Agency, U.S. *See* EPA.

EPA. 2003a. Frequently Asked Questions, Superfund Redevelopment Initiative. http:// www.epa.gov/superfund/ programs/recycle/communit/t_faqs.htm. Accessed 15 June 2003.

EPA. 2003b. SARA Overview. http://www.epa.gov/superfund/action/law/sara.htm. Accessed 27 May 2003.

EPA. 2003c. Superfund Sites Returned to Productive Use. http://www.epa.gov/superfund/ programs/recycle/success/ list170.htm. Accessed 12 March 2003.

Gupta, S., G. Van Houtven, and M. L. Cropper. 1995. "Do Benefits and Costs Matter in Environmental Regulation? An Analysis of EPA Decisions under Superfund." In Revesz and Stewart 1995a: 83-114.

Haimes, Y. Y. 2004. *Risk Modeling, Assessment, and Management.* New York: John Wiley & Sons, Inc.

Hamilton, J. T., and W. K. Viscusi. 1995. "The Magnitude and Policy Implications of Health Risks from Hazardous Waste Sites." In Revesz and Stewart 1995a: 55-82.

Hamilton, J. T., and W. K. Viscusi. 1999. "How Costly Is 'Clean'? An Analysis of the Benefits and Costs of Superfund Site Remediations. *Journal of Policy Analysis and Management* 18 (1): 2-27.

Hersh, R., K. N. Probst, K. Wernstedt, and J. Mazurek. 1997. "Linking Land Use and Superfund Cleanups: Uncharted Territory" Internet Edition. Washington,DC:. Resources for the Future. www.rff.org/reports/PDF_files/ landuse.pdf#land%20use

Hird, J. A. 1993. "Environmental Policy and Equity: The Case of Superfund." *Journal of Policy Analysis and Management* 12 (2): 323-343.

Hird, J. A. 1994. *Superfund: The Political Economy of Environmental Risk.* Baltimore: The Johns Hopkins University Press.

Kenney, M. A. 2001. "Development of a Value-Based Model to Provide Options for Reuse of Superfund Sites." Undergraduate thesis. University of Virginia.

Kornhauser, L. A., and R. L. Revesz. 1995a. "De Minimis Settlements under Superfund: An Empirical Study." In Revesz and Stewart 1995a: 187-218.

Kornhauser, L. A., and R. L. Revesz. 1995b. "Evaluating the Effects of Alternative Superfund Liability Rules." In Revesz and Stewart 1995a: 115-144.

Lambert, J. H., J. A. Baker, and K. D. Peterson. 2001. "Decision Aid for Allocation of Transportation Funds to Guardrails." *Accident Analysis and Prevention* 35: 47-57.

Ma, H., and D. J. Crawford-Brown. 1997-1998. "Comparison of Systems and Single-Medium Approaches in Environmental Risk-Based Decision-Making: A Case Study of a Sludge Management Problem." *Journal of Environmental Systems* 26 (3): 215-247.

Madu, C. N. 1999. "A Decision Support Framework for Environmental Planning in Developing Countries." *Journal of Environmental Planning and Management* 42 (3): 287-313.

Marshall, K. T., and R. M. Oliver. 1995. *Decision-making and Forecasting*. New York: McGraw-Hill, Inc.

Mazurek, J., and R. Hersh. 1997 (July). "Land Use and Remedy Selection: Experience from the Field – The Abex Site." Washington, DC: Resources for the Future. www.rff.org/CFDOCS/disc_papers/PDF_files/9726.pdf

Mishan, E. J. 1976. *Cost-Benefit Analysis*. New York: Praeger Publishers.

Powell, J. D. 1988. "A Hazardous Waste Site: The Case of Nyanza." In *Environmental Hazards: Communicating Risks as a Social Process*, Sheldon Krimsky and Alonzo Plough, eds., 239-297. Dover: Auburn House Publishing Company.

Probst, K. N. 1995. "Evaluating the Impact of Alternative Superfund Financing Schemes." In Revesz and Stewart 1995a: 145-170.

Revesz, R. L., and R. B. Stewart, eds. 1995a. *Analyzing Superfund: Economics, Science, and Law*. Washington, DC: Resources for the Future.

Revesz, R. L., and R. B. Stewart. 1995b. "The Superfund Debate." In Revesz and Stewart 1995a: 3-24.

Stewart, R. B. 1995. "Liability for Natural Resource Injury: Beyond Tort." In Revesz and Stewart 1995a: 219-250.

Stokey, E., and R. Zeckhauser. 1978. *A Primer for Policy Analysis*. New York: W. W. Norton & Company.

U.S. Congress, Office of Technology Assessment. 1985 (April). *Superfund Strategy*. OTA-ITE-252. Washington, DC: U.S. Government Printing Office.

U.S. Congress, Office of Technology Assessment. 1988 (June). *Are We Cleaning Up? 10 Superfund Case Studies*. Special Report. OTA-ITE-362. Washington, DC: U.S. Government Printing Office.

U.S. Department of Energy, Office of Environmental Guidance. 1994 (February). *The Hazard Ranking System (HRS)*. EH-231-024/0294.

U.S. House of Representatives. U.S. Code Title 42 – Chapter 103. http://uscode.house.gov/title_42.htm (Accessed 27 May 2003).

Walker, K. D., M. Sadowitz, and J. D. Graham. 1995. "Confronting Superfund Mythology: The Case of Risk Assessment and Management." In Revesz and Stewart 1995a: 25-54.

Wernstedt, K., and K. N. Probst. 1997 (July). *Land Use and Remedy Selection: Experience from the Field – The Industri-Plex Site*. Washington, DC: Resources for the Future. www.rff.org/CFDOCS/disc_papers/PDF_files/9727.pdf

Wernstedt, K., R. Hersh, and K. N. Probst. 1988 (March). *Basing Superfund Cleanups on Future Land Uses: Promising Remedy or Dubious Nostrum?* Washington, DC: Resources for the Future. www.rff.org/CFDOCS/disc_papers/ PDF_files/9803.pdf

5

A COST-BENEFIT MODEL FOR EVALUATING REMEDIATION ALTERNATIVES AT SUPERFUND SITES INCORPORATING THE VALUE OF ECOSYSTEM SERVICES

Melissa Kenney
Duke University Nicholas School of the Environment

Mark White
University of Virginia McIntire School of Commerce

INTRODUCTION

In 1980, Congress passed the Comprehensive Environmental Response, Compensation, and Liability Act (CERCLA) in response to a particularly unfortunate incident in the Love Canal area of Niagara Falls, New York, in which numerous schoolchildren were exposed to toxic chemicals from an abandoned waste disposal site. The U.S. Environmental Protection Agency (EPA) was charged with establishing, administering, and enforcing policies and procedures through which the nation's worst hazardous waste sites (i.e., those posing the greatest risks to human health) might be identified, remediated, and returned to productive use. Further, the Act established an endowment, nicknamed "Superfund," to assist with cleanup costs and imposed substantial liability on owners, transporters, and generators of hazardous waste materials.

Despite a quarter-century of experience and expenditures of more than $28 billion, the Superfund program has enjoyed a mixed record of success. Since its inception, more than 1500 sites have been identified and added to the National Priority List (NPL), but only about 275 sites have been completely cleaned up and deleted from the NPL. Of the deleted Superfund sites, approximately 240 sites have been returned to productive use (USEPA 2003, 2004, 2005). A number of factors, including uncertainties regarding the extent and severity of contamination, efficacy of remediation technologies, costs, community pressures, etc. have hindered the Agency's efforts to facilitate more expeditious cleanups (Probst et al. 2001; Probst & Sherman 2004).

Remediation and reuse of Superfund sites has also been slowed by managerial impediments. After a potentially hazardous site is brought to EPA's attention, cleanup actions are implemented using a fairly well-defined procedure. The cleanup process begins with a preliminary assessment/site inspection (PA/SI). If the situation is serious and warrants remediation, the site is listed on the NPL and a Remedial Investigation/ Feasibility Study (RI/FS) is initiated to determine the nature and extent of contamination. After analyzing these findings, EPA officials issue a Record of Decision (ROD) identifying the cleanup alternatives to be used in remediating the site. The Remedial Design/Remedial Action (RD/RA) plan follows, stipulating the actual technologies and standards to be attained. Successful site cleanups are designated "Construction Complete," after which they pass into the Post Construction Completion phase involving monitoring, remedy optimization, and eventual deletion from the NPL site registry.

However, numerous individuals and organizations, including the EPA itself, have criticized this process as inflexible, inappropriate, inefficient, and insensitive to the needs of surrounding communities (Schneider 1993). In 1993, then EPA administrator Carol Browner initiated reforms aimed at making the program work faster, fairer, and more efficiently, and since that time, improvements have been made.

Yet despite these reforms, more could still be done. Cannon (2006) argues for the adoption of an adaptive management model in administering Superfund sites. He identifies three major challenges to EPA's current operating procedures. First, he suggests that the Environmental Protection Agency's model of a time-limited intervention is both unrealistic and untenable given the complexities and uncertainties surrounding cleanup efforts at hazardous waste sites. Second, he takes issue with the Agency's narrow focus on protecting public health. This suggests that a broader interpretation of "protectiveness," incorporating additional ecological, economic, and social values, is called for under the statutes. Finally, Cannon notes that although the Federal government has non-delegable authority for selecting a cleanup remedy, a site's future reuse is ultimately the result of many decisions made by private parties, municipalities, and states.

An adaptive management approach is well suited for addressing each of these concerns and shows promise for improving the management of Superfund sites (Cannon 2006). Implementation, of course, will not be easy, and will require substantial flexibility and patience on the part of Superfund administrators, site managers, potentially responsible parties (PRPs), and community leaders. It will also require new models and methods of analysis for evaluating tradeoffs between ecological, economic, and human health concerns.

ORGANIZATION OF THIS CHAPTER

This chapter describes an effort to model the selection of remediation alternatives as part of an EPA grant awarded to researchers at the University of Virginia in 2001-2004. While standard decision-making theory urges modelers to "begin with the end in mind," proponents of adaptive management encourage flexibility and an openness to change as organizational learning raises the possibility of better end-use alternatives and remediation techniques. The framework described in this chapter, which is adaptable in both time and space, fulfills both conditions.

The backbone of our model is a standard discounted cost-benefit analysis of alternative remediation technologies. Allowing for the possibility of different end-uses makes the model more interesting from an adaptive management perspective, and it becomes even more intriguing when values for ecosystem services are included in the ecological reuse scenario. Our model was parameterized using data from Emmell's Septic Landfill, an actual Superfund site in New Jersey. Analysis of the model's results using Monte Carlo techniques identified a wide discrepancy in preferred treatment regimes and highlighted critical decision elements.

In the first section of this chapter we provide some background discussion on several key aspects of our model – benefit-cost analysis, remediation alternatives, and ecosystem services. Then we describe the model's general structure, highlighting the benefits of reusing Superfund sites, identifying cost categories, and describing treatment alternatives. After a brief section describing the data used in analyzing our specific Superfund site, we present our results. The chapter concludes with a discussion of these results and recommendations for further study.

BACKGROUND

Benefit-Cost Analysis

Decision scientists are fond of saying, "All models are wrong, but some are useful." Numerous alternative models could be used in analyzing tradeoffs between remediation methods and reuse opportunities, but we employed a benefit-cost framework for several reasons.

First, benefit-cost models allow for explicit consideration of tradeoffs between and among competing uses and competing remediation technologies. As with any endeavor, resources (financial, managerial, environmental, etc.) are assumed to be in limited supply and subject to alternative uses. When markets are well-functioning, maximizing the difference between benefits and costs also maximizes societal welfare.

Second, benefit-cost models are relatively simple to construct, easy to understand, and informed by many sources. In our model, the selection of remediation alternatives might be based on scientific judgments (e.g., contaminant types and locations, available technologies), while reuse opportunities could arise from stakeholder preferences (e.g., commercial opportunities and community needs).

Third, by combining heterogeneous preferences into a single metric, benefit-cost analysis provides a method for reaching a decision that incorporates all views. Builders, for instance, might prefer to see a waste site redeveloped into residential housing lots, while birdwatchers might value a reforestation project more highly, based upon what each group stands to gain individually. Properly done, a benefit-cost analysis accounts for both of these perspectives.

Finally, benefit-cost analysis is required or strongly encouraged by numerous federal regulations, and it would be folly to presume that financial considerations are absent from the deliberation of remediation alternatives (Farrow & Toman 1999). Executive Order 13258 (issued in 1996) requires the benefits of Federal regulation to justify their costs.[1]

One could claim that our model oversimplifies a complex decision-making process by using profit as the sole decision-making criterion. This is true. Multi-attribute models – in which alternatives are evaluated against multiple objectives – can capture additional intricacies, albeit at the cost of increased complexity (Keeney & Raiffa 1993; Merkhofer, Conway & Anderson 1997), and in fact, Merkhofer et al. (1997) use a multi-attribute model to assess the siting of a hazardous waste management facility. However, it is generally not feasible to conduct both a benefit-cost analysis and a multi-attribute utility analysis at the same site due to limited financial resources, and the former is generally less expensive (by virtue of including fewer objectives, if for no other reason!). Moreover, decision-makers will often *implicitly* combine the results of a benefit-cost analysis with other objectives in reaching a remediation and reuse determination (Lange & McNeil 2004).

Remediation Alternatives

Superfund sites are evaluated individually to determine which remediation alternative would best clean up the site. The remediation method chosen is the one that most quickly restores the site to its pre-contaminated

---

[1] This order superseded earlier mandates, e.g., Executive Orders 12291 and 12866, requiring strict cost-benefit analyses of Federal regulations.

condition. Remediation decisions generally do not consider what the site might be used for after it is cleaned up.

An alternative to current practice would encourage site managers to consider remediation strategies in the context of likely site reuses. Thinking about future reuse before remediation occurs could increase the probability of site reuse, reduce remediation and reuse costs, and allow for a more efficient allocation of limited resources.

Ecosystem Services

The world's ecosystems provide numerous benefits to humanity. These include both traditional natural resources (e.g., forests, fisheries, agricultural products) and various important life-support functions, collectively termed "ecosystem services" (Daily 1997; National Research Council 2005). Interest in the valuation and preservation of ecosystem services has grown as human populations and consumption levels have increased. Seemingly inexhaustible resources are now scarce, and current accounting methods fail to provide the necessary economic signals allowing efficient allocation of remaining supplies (Chichilnisky & Heal 1998; Costanza 2000; Heal 2000).

There is not yet (and may never be) any exhaustive taxonomy of ecosystem services. Drawing upon the work of de Groot et al. (2002), the United Nations' Millennium Ecosystem Assessment program recognizes four broad categories of services. *Provisioning services* deal with the provision of natural resources, e.g., food, fibers, water, wood, medicines, and the like. *Regulating services* refer to the maintenance of essential ecological processes and life-support systems, e.g., climate, flood protection, and water quality. *Supporting services* include soil formation, nutrient cycling, and habitats for wild plant and animal species. *Cultural services* provide recreational, aesthetic, and spiritual benefits (Millennium Ecosystem Assessment 2005). Our model estimated values for timber, carbon sequestration, water quality improvement, hunting, nonconsumptive wildlife use, and aesthetics.

Ecosystem services have historically been undervalued. In an important article in *Nature*, Costanza et al. (1997) attempted to estimate the economic value of all Earth's life-support functions. The authors arrived at an annual figure of US$ 33 trillion for 17 ecosystem services performed across 16 biomes, noting that by comparison, the world's gross national product (the sum of goods and services produced by all humankind) was only US$ 18 trillion.[2]

---

[2] Costanza and his coauthors (Costanza et al. 1997) have been criticized on both theoretical and methodological grounds (El Serafy 1998; Pearce 1998; Sagoff 1997; Toman 1998). Their paper nonetheless remains an important contribution for raising consciousness about the importance of ecosystem services.

In valuing ecosystem services, economists make a distinction between *use* and *non-use* values. Use values may be further differentiated into consumptive and nonconsumptive uses. The former includes fisheries, timber, and wild berry harvesting. Examples of the latter are birdwatching, swimming, and pollination services performed by wild bees and moths. Nonuse values arise from people's willingness to pay for things' values unrelated to the actual or intended use of a particular ecosystem component, e.g., the mere knowledge that certain charismatic species still exist. Numerous techniques, e.g., hedonic pricing, travel costs, and contingent valuation, are available for estimating these values, and an extensive literature has arisen delineating best practices (Arrow et al. 1993; Bateman & Willis 2001; Champ, Brown & Boyle 2004; Cummings, Brookshire & Schulze 1986).

Nonetheless, the identification and valuation of ecosystem services is fraught with uncertainties. Ecologists find it difficult to delineate system boundaries, quantify ecosystem functions, and aggregate across trophic levels. Economists are challenged by a lack of markets for most ecosystem services, methodological dilemmas, and the very notion of valuing what are, essentially, invaluable services (Sagoff 1997). Although there are several repositories of studies from which one can glean inferences and make comparisons, e.g., the Ecovalue Project (Gund Institute for Ecological Economics 2005) and the Environmental Valuation Reference Inventory (Environment Canada 2005), a key issue is the degree to which results obtained from one study site can be extrapolated to other locations. Our results, which are based on published and unpublished reports, are subject to all of these uncertainties.

Previous Research

Although relatively few models of remediation alternatives at hazardous waste sites appear in the literature, one can be sure many more are in use at actual Superfund sites. The following studies relate most closely to the work at hand.

Smedes et al. (1993) report a successful remediation action at a former tire manufacturing plant in California. Activated-charcoal air strippers were used to remove and vaporize volatile organic compounds extracted from groundwater over a period of seven years at a cost of $22 million. The authors found a close match between predicted and actual performance of the pump-and-treat remediation method.

Toland, Kloeber, and Jackson (1998) modeled remediation alternatives for contaminated nuclear-processing sites supervised by the Department of Energy (DOE). As in the present work, they employed a discounted benefit-cost framework to select among several different treatment alternatives. They also used Monte Carlo analysis to analyze their work, and found a tradeoff between cost- and time-effectiveness. An important discovery was that monitoring costs, which of course are quite important for nuclear waste sites, can have a significant impact upon total life-cycle costs. Also using data from the DOE, Jones et al. (2000) found cleanup initiatives at nuclear waste sites to be quite sensitive to the program's overall *goals*. Perhaps not surprisingly, different objectives (e.g., reduction of current-period vs. end-of-period risks) resulted in different amounts of waste treated, costs, and risks reduced over a ten-year period.

Kiel and Zabel (2001) reported encouraging results in their study of cleanup benefits at two Superfund sites in Woburn, Massachusetts. Using hedonic pricing methods, they found an expected increase in residential property values following remediation activity. Despite $70 million estimates of cleanup costs – at each site – the researchers found benefits in the range of $72 to $122 million, concluding that "it is likely that these benefits are greater than the present value of the estimated costs of cleaning up these sites."[3] Farber (1998) reviews additional studies on the effects of hazardous waste sites on nearby property values.

Finally, important work by Hamilton and Viscusi highlighted the uncertainties involved in performing these kinds of analyses, noting the difficulties of summing effects from multiple pathways and differences arising from actual vs. perceived risk assessments (Hamilton & Viscusi 1999; Viscusi & Hamilton 1999; Viscusi, Hamilton & Dockins 1997). Most tellingly, they found cleanup efforts at most Superfund sites would not pass a benefit-cost test if actual risks of exposure were evaluated against cleanup costs (Hamilton & Viscusi 1999).

OVERVIEW OF THE MODEL

Our model, hereafter the Superfund Adaptive Reuse and Redevelopment (SARR) model, was developed to analyze remediation alternatives at a Superfund site, assuming multiple possible end-uses, and accounting for the value of ecosystem services. In creating this model, we employed Herendeen's (1998) LIEDQO environmental analysis framework in classifying the economic implications of remedy choice and reuse alternatives. According to Herendeen, quantitative analysis of environmental problems should entail consideration of six key elements:

---

[3] Probst et al. (1995) report the average cleanup cost at a Superfund site to be $30 million.

L (limits) ... e.g., cleanup standards for various pollutants

I (indirect effects) ... e.g., aesthetic benefits associated with phytoremediation

E (efficiency) ... e.g., quantification of linkages between treatment time and eventual remediation

D (dynamics and lags) ... e.g., the use of discount rates to equate future benefits with present remediation costs

Q (equity) ... e.g., relative distribution of costs and benefits – our model does not explicitly address this issue

O (other) ... e.g., uncertainty – addressed using Monte Carlo techniques

Careful consideration of each of these six LIEDQO elements helped to ensure the model's completeness. An overview of our model is shown in Figure 5.1.

Because alternatives for reuse are closely tied to the specifics of a particular Superfund site, it is difficult to make objective generalizations about appropriate reuse alternatives for a given site. We used subjective judgment in arriving at two reuse options (residential and ecological) for our study site.

Three categories of costs were recognized: external costs, remediation costs, and site-specific costs. External costs included mortality and morbidity costs resulting from the presence of toxic wastes. Remediation costs referred to the direct costs of cleanup under each of the four treatment regimes. Reuse-specific costs reflected additional costs resulting from a particular reuse scenario. Future benefits and costs were estimated in current dollars, adjusted for inflation, and then discounted back to compute net present values for each of the eight alternative scenarios.[4]

Economists make use of net present values to aggregate benefits and costs occurring across time. Discounting reflects the fact that money has time value, i.e., most people would choose to receive $1 today vs. receiving the same amount at some time in the future. The discount rate reflects one's preference between the two sums – the higher the rate, the more one prefers current over future consumption.

The SARR model's overall structure is quite flexible, and capable of handling more or fewer reuse alternatives and remediation treatments than the four regimes determined as appropriate for our site. We applied the model in

---

[4] Two reuse alternatives x four treatment regimes = eight remediation-reuse scenarios.

its most flexible form, i.e., to identify the *combination* of reuse and remediation alternatives yielding the highest net present value. However, a slight reframing of our model would allow decision-makers to discern the best remediation strategy given a specific reuse alternative, or conversely, the best reuse alternative given a preferred treatment regime. Whichever approach one selects, the decision criterion should always be to choose the alternative yielding the highest net present value.

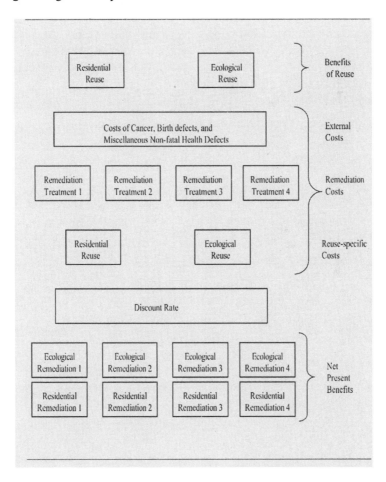

Figure 5.1. Conceptual diagram of the SARR model.

APPLICATION OF THE MODEL

Description of the Study Site

After developing the SARR model, we applied it to Emmell's Septic Landfill, a 38-acre Superfund site located in a suburb of Atlantic City, New Jersey. The landfill previously operated as a repository for septic and sewage sludge. In violation of its permit, it also accepted household garbage, tires, paint sludge, gas cylinders, and construction/industrial wastes. As a result, the soil and groundwater at this location became contaminated with a combination of volatile organic compounds (VOCs) and heavy metals. The site was added to the National Priority List in 1999, and since that time, an emergency removal action at the site has eliminated the most highly contaminated soil (USEPA 1999).

Model Parameterization

Appropriate parameterization is one of the most difficult tasks in modeling. Parameterization is difficult because reasonable data may be difficult or impossible to locate, or because what data is available is subject to large amounts of uncertainty. There are three major sources of data for model parameterization: (1) Sample data, (2) Literature sources, and (3) Subjective judgments.

Our study relied on data drawn from published and unpublished sources and some subjective judgments by experts in the field. In applying this information to a particular study site, consideration must be given to its transferability. For example, Salzman, Thompson, and Daily (2001) estimated the value of watershed services for New York City. Because our study site is quite close to New York City and subject to similar cultural demands, it may be appropriate to use these figures as estimates for the value of watershed services in New Jersey. However, it might be very inappropriate to utilize numbers derived from watershed valuations performed in Indonesia.

The use of subjective judgments also deserves mention. Best practice in the use of subjective judgments is to solicit experts' opinions and to make these judgments explicit (Meyer & Lyons 2000; Morgan & Henrion 1990). Multiple judgments can also be used, thereby explicitly incorporating a range of judgments and the associated uncertainty of the parameter value (Clemen & Reilly 2001). For example, if one needed a value for watershed services and it was not available through sampling or in the literature, one might ask a group of knowledgeable experts to estimate the median and range of monetary benefits for watershed services at a particular site. This information could then be aggregated into probability distributions representing current knowledge about watershed service values at a particular site.

Variable Selection

All models are extrapolations of reality, and modelers must walk a fine line between complexity and authenticity. We believe the choices we made in parameterizing our model of remediation-reuse alternatives for Emmell's Septic landfill were realistic without being unduly simplistic. As noted earlier, we identified two options for reuse and four alternative treatment regimes. In solving the model, values for 118 different variables were estimated. Forty-eight of these were associated with external costs, including cancer costs (5), birth defects (16), and miscellaneous health conditions (17). Remediation costs accounted for another 22 variables: five each for the air stripper, surfactant injection, and phytoremediation treatment methods, and seven related to the excavation alternative. We estimated 19 site-specific variables linked to ecological reuse and 40 variables associated with the development and eventual sale of residential lots.

Reuse Alternatives

Decision-makers at Superfund sites generally consider only a subset of all possible reuse options, focusing on those most compatible with the site's immediate surroundings. Based on the site's geography, zoning, and location near other residential housing developments, redevelopment into residential lots seemed a reasonable reuse alternative (Bonnani 2002). The site's location within New Jersey's pine barrens, an environmentally sensitive habitat (Forman 1979), together with recent interest in the value of ecosystem services, prompted identification of our second reuse alternative as a recreational woodlot.

Benefits of Residential Reuse

We approximated the benefits of residential reuse using the value of estimated lot sales from the redeveloped property. In so doing, we knowingly set boundaries on our analysis, excluding, for example, additional sales at local businesses due to a slightly larger population, increased tax revenues, and the like. We also excluded increased costs arising from a greater residential population (e.g., traffic, police services, schooling). In both cases we reasoned that the Emmell's site was but one housing alternative in the area, and that if it were not developed, another property would be. Our calculations assumed the creation of seven lots out of the 38-acre parcel.

Benefits of Ecological Reuse

The benefits of ecological reuse arose from the ecosystem services provided if the landfill were cleaned up and allowed to revert to a recreational woodlot. Our model considered several benefits of ecological reuse, including carbon sequestration services, hunting, nonconsumptive wildlife use (e.g., birdwatching), aesthetics, and water purification services.

Increased levels of carbon dioxide (CO_2) are believed to be partly responsible for global warming and climate change. Numerous efforts to reduce these emissions are afoot, and various markets have arisen facilitating trades in CO_2 emissions, which allow pricing of carbon sequestration services. For our study site, we assumed an annual sequestration rate of 28 tons per acre and a market price of $2 per ton (Dolman, Moors & Elbers 2002; Huang & Konrad 2001; Newell & Stavins 2000).

The annual value of hunting, wildlife, and aesthetics services was extrapolated from published studies by assuming a usage rate of five people per day for the entire year (with the exception of hunting, which lasted only one quarter of the year) and an average daily per-person willingness-to-pay of $45.05, $43.06, and $10.53 respectively (Cummings, Brookshire & Schulze 1986; Goodwin, Offenbach, Cable & Cook 1993; Willis 1989). The annual value of water purification services was estimated from Salzman, Thompson & Daily's (2001) report on the Catskills watershed at $1,960 per acre.

External Costs

A major reason for remediating contaminated sites is to reduce the risk of human exposure to hazardous substances, in accordance with EPA's mandate to protect human and environmental health. Unless harmful pollutants are removed or contained, they pose a risk to human health, imposing both direct and indirect costs on society (Johnson & De Rosa 1997). In our model these increased health risks and associated medical costs are referred to as external costs.

External costs can be proxied by the direct medical costs and productivity losses arising from health impairments resulting from environmental contamination. Among the external costs considered by our model were the occurrence of excess cancers, birth defects, and miscellaneous health conditions in the exposed population.

External costs are particularly difficult to quantify for three reasons: (1) Effects are usually determined for a single pollutant and not multiple pollutants; (2) Exposure risk differs due to site-specific characteristics; and (3) Quantifying the risk and cost of the increased risk at a site or group of sites involves a somewhat haphazard method of guess-work.

Most research on the effects of hazardous pollutants on humans is based on exposure to a single contaminant. Almost all pollutant problems at Superfund sites, however, result from multiple contaminants, sometimes known as "chemical soup." Although human health problems have been linked to Superfund site contamination (Geschwind et al. 1992; Sosniak, Kaye & Gomez 1994), it is difficult to predict the overall magnitude and scope of the health response, as the impacts are not necessarily additive.

In addition to uncertainties surrounding the dose-response relationship, there are risks associated with actual human exposure to contaminants. Populations are subject to different levels of exposure depending on the contaminant type and location. For example, groundwater contamination poses a much greater exposure risk to a rural population relying on wells for its drinking water than it does to city-dwellers served by a municipal water supply.

One of the reasons it is so difficult to quantify contaminant risk and exposure is that there is little data available. Despite reasonable evidence demonstrating the adverse effects of increased levels of pollutants on an exposed population, it is difficult to link this evidence to the actual practice of public health (Johnson & De Rosa 1999). As a result, it is hard to quantify the magnitude in medical costs of pollutant exposure.

Despite the difficulty of quantifying external costs due to pollutant risk, it is well accepted that some exposure to pollutants increases a population's risk of poor health effects and thus health-related costs. For example, Lybarger et al. (1999) found that exposure to volatile organic compounds (VOCs) led to an increase in medical costs for the exposed population. The authors acknowledged that while it was impossible to determine medical costs precisely, it was possible to obtain estimates of their general magnitude. Their analysis demonstrates current "best practices" in this challenging area of valuation.

Our model was parameterized using values drawn from the literature and expert judgment. The values we used are shown in Figure 5.2; additional information is available from the authors upon request.

Remediation Costs

It is often the case at Superfund sites that pollutants will infiltrate multiple media, requiring multiple treatment alternatives to completely address the situation. At Emmell's Septic Landfill, the groundwater is contaminated by volatile organic carbon solvents and heavy metals, while the soil contains heavy metals and other toxic substances. An effective treatment regime must address both problems. We considered two alternatives for groundwater remediation, and two options for treating soil contamination.

A. Birth Defects			B. Health Effects		
Birth rate	0.0142		Susceptible population	25.0%	
Population	9,334				
Type of Defect	**Excess Cases**	**Cost per case**	**Health Effect**	**Excess Cases**	**Cost per case**
Cardiovascular	0.0047	$ 356,000	Anemia	0.0160	$ 390
Central nervous system	0.0009	350,000	Diabetes	0.0160	959
Chromosonal	0.0020	537,000	Eczema & skin disease	0.0120	178
Cleft lip and palate	0.0017	120,000	Stroke	0.0050	7,800
Gastrointestinal	0.0022	173,000	Urinary tract disease	0.0220	424
Genitourinary	0.0011	158,000	Hearing impairment	0.0220	446
Musculoskeletal	0.0012	157,000	Speech impairment	0.0260	446
			Renal disease	0.0350	9,050

C. Cancer Costs

Affected Population	Excess Cancer Risk	Probability of Survival	Productivity Loss	Medical Costs	Cost per case (death)
21,281	0.00011	0.59	$ 13,420	$ 51,400	$ 6,070,000

Sources: (E² Inc., 2002)

Figure 5.2. External costs.

Groundwater Treatment Alternatives

The most common method for remediating contaminated groundwater is "pump and treat" (USEPA 1994). In this technique, groundwater is pumped to the surface, its volatile contaminants are vaporized using an air stripper, and the resulting purified stream is disposed of in the storm-water system (National Research Council 1999; USEPA 1994). The second method we considered employs the same method (e.g., pump-and-treat using an air stripper), but adds the injection of a chemical surfactant. Injecting a surfactant into the subsurface accelerates the release of contaminants from the soil surrounding the aquifer into the groundwater (Fountain 1997; National Research Council 1999). The result is a more rapid remediation (Fountain 1997).

Alternatives for Soil Remediation

The alternatives for soil remediation differ dramatically from one another in the time required to effect a complete remediation. The first method, *phytoremediation*, is a passive process in which natural or introduced vegetation concentrates and removes soil contaminants over a long period of time (National Research Council 1999). The second method involves excavation and removal of the contaminated soil, with subsequent offsite treatment or disposal in a hazardous waste landfill (National Research Council 1999; USEPA 1994).

Treatment Regimes and Costs

Combining the need for both groundwater and soil remediation with two treatment alternatives for each medium results in the four treatment regimes shown in Figure 5.3.

		Soil Remediation Method	
		Phytoremediation	Excavation
Groundwater Remediation Method	Air Stripper	Treatment 1	Treatment 2
	Air Stripper with Surfactant	Treatment 3	Treatment 4

Figure 5.3. Treatment regimes for Emmell's Septic Landfill.

Clearly, these regimes differ with respect to both cost and time. Phytoremediation combined with a "pump and treat" protocol, for example, has the lowest remediation cost, but takes a long time. Excavation and use of an air stripper is quick, but expensive. Overall, the cost of a particular treatment regime must take into account the initial equipment outlay, together with any ongoing operating expenses. Initial costs include equipment rental, labor, and equipment and materials. Ongoing operating costs include maintenance, treatment, and monitoring and reporting expenses. Estimates for the necessary cleanup of Emmell's Septic Landfill were obtained from several sources and are shown in Figure 5.4.

	Air Stripper	Surfactant	Phyto-remediation	Excavation
Rental Equipment	$ 121,900	$ 135,400	$ 31,190	$ 5,620,000
Labor	41,940	43,810	33,300	61,010
Equipment & Materials	102	102	71,880	na
Treatment	402,300	452,800	1,013	na
Monitoring & Reporting	85,720	85,720	na	na

Sources: R. S. Means and Company (2002a; , 2002b)

Figure 5.4. Groundwater and soil remediation costs.

Reuse-Specific Costs

To promote reuse of a Superfund site, it is generally necessary to prepare the site in some fashion after the environmental contamination has been removed or mitigated. Reuse-specific costs are those incurred for redeveloping the site for a particular reuse. We estimated reuse-specific costs for two reuse alternatives: ecological and residential reuse.

The cost of ecological reuse is the cost to restore the ecosystem functions of the site. We determined that the most likely ecological restoration for our site would be to revert the site back to a pine barren. Therefore, the cost to restore the ecological function to the site would be purchasing and replanting appropriate vegetation after site remediation.

The costs associated with residential reuse were more complex to estimate. Based on surrounding land-use patterns, we determined the most likely commercial use of this site would be as residential housing. Thus, we estimated expenses needed to develop the land for sale as housing lots. We considered four categories of development costs: fees and permits, earthwork, water and sewage, and infrastructure costs. Reuse-specific costs are summarized in Figure 5.5.

Inflation and Discount Rates

We controlled for inflation by assuming costs and benefits increased at an inflation rate of 2 percent per year – a conservative estimate given Ibbotson and Sinquefeld's long-run empirical finding of 3.1 percent per year (Ibbotson Associates 2005). Determining an appropriate discount rate posed a bigger problem. Discounting is clearly established in both financial theory and practice, but it poses some challenges with regard to environmental

Residential Reuse*

Fees and permits		Water and Sewage	
Number of lots	7	Water main	750 ft
Municipal fees	$ 30,000	Water main cost	$ 20/ft
Zoning fee	2,500	Water service connection cost	$ 750/lot
Plan review fee	500	Contingency percentage	10 %
Bond and letter of credit	300	RCP	300
		RCP Cost	22.5
Earthwork		Structures	3 structures
Earthmoving volume	3,200 cu yds	Cost of structures	$ 2,800 each
Earthmoving cost	$ 2/cu yd	Storm water pond	$ 14,500
Fill work volume	3,200 cu yds	Rip rap	60 feet
Fill work cost	$ 2/cu yd	Rip rap cost	$ 65
Rough grade	3,200 cu yds	Contingency percentage	10 %
Rough grade cost	$ .75/cu yd		
Contingency percentage	20 %	**Infrastructure**	
		Lighting	$ 3,500
		Landscaping	5,000
		Area to be paved	3,150 feet
		Subgrade	150
		Subgrade preparation cost	8
		Base stone cost	8
		Base paving cost	6
		Surface paving cost	3.75
		Contingency percentage	15 %
		Vegetation	
		Seeding	3500
		Acres to seed	8

Ecological Reuse*

Vegetation Seeding	$3500/ acre
Acres to Seed	8 acres

Sources: Neill (2002), R. S. Means and Company (2002a; , 2002b)

Figure 5.5. Reuse-specific costs: Residential reuse and ecological reuse.

sustainability, for it contains a built-in bias against future generations, i.e., the higher the discount rate, the faster resources are likely to be depleted (Goulder & Stavins 2002; Pearce & Turner 1990; Rabl 1996). Arguments can be made for both low social rates of discount and higher, market-based rates. In the end, we compromised and used EPA's recommended annual discount rate of 7 percent.

Monte Carlo Analysis

Monte Carlo analysis is a mathematical tool for computing the probability a particular outcome will be realized given uncertainties in the input variables. Named after the famed casino gambling city in Monaco, it was developed by John von Neumann and Stanislaw Ulam while working on the atomic bomb during World War II.

Monte Carlo analysis employs a random selection process similar to that present in games of chance (Clemen & Reilly 2001; Morgan & Henrion 1990). The analysis simulates a large number of possible outcomes by generating probability distributions for each uncertain variable and then computing through with random "draws" from each variable's respective probability universe. After performing a large number of trials, a final result is obtained, also in the form of a probability distribution (Clemen & Reilly 2001; Morgan & Henrion 1990).[5]

RESULTS

Our results for the Emmell's Septic Landfill site were obtained using the methods described earlier and the values presented in the section on model parameterization. The outcome of our analysis was a set of eight net present values corresponding to the eight remediation-reuse scenarios. The mean values and standard deviations for each of these alternatives are presented in Figure 5.6.

All of the imagined reuse-remediation alternatives resulted in negative net present values, i.e., it is costly to clean up the site. The NPVs of the ecological reuse alternatives ranged from –$37.6 million to –$44 million. Net present values for the residential reuse alternative ranged from –$41.2 million to –$47.6 million. However, regardless of treatment type, ecological reuse dominated residential reuse, i.e., it resulted in *less* negative NPVs. Treatment 4 (air stripper with surfactant and excavation) had the best net present value

---

[5] 1,000 to 10,000 trials are not uncommon. Naturally, a computer is needed to perform this analysis. In practice, the number of trials depends on the use to which the forecast will be put and the availability of computing resources.

	Treatment 1	*Treatment 2*	*Treatment 3*	*Treatment 4*
	Air Stripper and Phyto-remediation	**Air Stripper and Excavation**	**Air Stripper with Surfactant and Phyto-remediation**	**Air Stripper with Surfactant and Excavation**
Ecological Reuse	– $ 42.1 million	– $ 44.0 million	– $ 41.8 million	**– $ 37.6 million**
	($10.8 million)	($9.5 million)	($10.8 million)	($7.7 million)
Residential Reuse	– $ 45.7 million	– $ 47.6 million	– $ 45.4 million	– $ 41.2 million
	($11.0 million)	($9.8 million)	($11.0 million)	($7.9 million)

*Results of Monte Carlo analyses with 10,000 trials; median values. Standard deviations in parentheses

Figure 5.6. Net present value of alternative reuses by treatment type.*

for both reuse alternatives. Treatment 2 (air stripper and excavation) had the worst NPV for both reuse alternatives.

The optimal treatment régime depends critically on which remediation strategy is chosen for both groundwater remediation and soil remediation. For groundwater remediation, surfactant injection is preferred over air strippers. For soil remediation, the optimal remediation depends on the groundwater remediation strategy. If air strippers are used, phytoremediation is slightly preferred over excavation. If an air stripper with surfactants is used, excavation is slightly preferred over phytoremediation.

Our model indicates that Treatment 4 (excavation and surfactant injection) and an ecological reuse scenario would likely result in the lowest net cost of remediation, i.e., the smallest monetary loss. Because the magnitudes for all of the remediation-reuse alternatives are quite similar, it seems likely that extrafinancial issues of the sort described in other chapters of this book will likely carry the day, emphasizing once more the need for a collaborative and adaptive management approach.

Tornado Analysis

Tornado charts are graphical representations of N-way data tables used in assessing the impact of each input variable *one at a time* on a particular forecast (outcome) variable and in displaying critical assumptions in descending order of importance. The resulting graph resembles the funnel-shaped cloud of a tornado. The results of running this procedure for Treatment 1 (air stripper and phytoremediation) in the ecological reuse scenario are presented in Panel I of Figure 5.7.

Panel I. Ecological Reuse – Treatment 1 (Air Stripper and Phytoremediation)

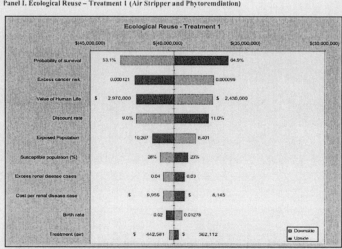

Panel II. Tornado Analysis Results: Key Variables in Rank Order (By Treatment Type)

Variable	Ecological Reuse				Residential Reuse			
	#1	#2	#3	#4	#1	#2	#3	#4
Probability of Survival	1	1	1	1	1	1	1	1
Excess cancer risk	2	2	2	2	3	2	3	2
Value of human life	3	3	3	3	4	3	4	3
Discount rate	4	4	4	5	2	4	2	4
Exposed population	5	5	5	4	5	5	5	5
Susceptible population (%)	6	6	6	6	6	6	6	6
Excess renal disease cases	7	7	7	8	7	7	7	8
Cost per renal disease case	8	8	8	9	8	8	8	9
Birth rate	9	10	9	10	9	10	9	10
Treatment cost (air stripper)	10				10			
Rental equipment (excavation)		9		7		9		7
Treatment cost (surfactant)			10				10	

Figure 5.7. Sample tornado analysis.

The output from the tornado chart is quite easily interpreted. Each bar in a tornado chart represents the impact (on the output variable) of varying a particular input variable by the same amount, e.g., -10% or +10%. The higher a variable's position in the chart, the greater its influence on the net present value of a particular remediation-reuse alternative. Probability of survival is the major variable driving value in each of our eight remediation-reuse scenarios (Panel II of Figure 5.7).

Dark grey bars correspond to an *increase* in a given input variable, while light grey bars represent *decreases*. For input variables that are positively correlated with the output variable (e.g., probability of survival and NPV of remediation-reuse), the dark grey bar will lie on the right and the light grey bar will lie on the left. For negatively correlated variables (e.g., cancer risk), the situation is reversed.[6] The maximum and minimum values associated with a specified change in a given input variable are shown at the outer limits of each bar.

Tornado charts prepared for each of the various remediation-reuse scenarios were extremely similar to one another, as is evident from Panel II. Probability of survival, excess cancer risk, value of a human life, discount rate, and population size were the top five value drivers in each case. Discount rates took on somewhat greater importance in the residential reuse scenario under Treatments #1 and #3, probably due to the longer times associated with phytoremediation.

Note that tornado charts examine each input variable independent of the others. This process is sometimes called "one-at-a-time perturbation" or "parametric analysis." It is useful for measuring the sensitivity of an output variable relative to a set of input variables or to quickly prescreen variables to assess which ones deserve further scrutiny, but depends quite heavily on the base case used in the analysis. Also, the results shown in a tornado chart do not allow for correlations *between* the input variables.

DISCUSSION

Comparison with Other Studies

It is difficult to compare the results of our study to results from previous studies, as many of the latter did not consider the range of parameters that we chose to include in the SARR model, and our results are *net* present values – they also include the *benefits* of reuse alternatives. We will, however, attempt to put the results of our research in the context of these studies.

Probst et al. (1995) reported average cleanup costs for a Superfund site to be $30 million, which is in line with our estimates for this project. In contrast, the Emmell's Septic Landfill RI/FS determined the cost of a pump-and-treat system at $4.9 million over an interim period of five years. Our cost estimate is substantially higher than this estimate, partly because we accounted for costs over the entire remediation period, but primarily due to our inclusion of external costs.

Kiel and Zabel (2001) obtained a positive NPV in their benefit-cost analysis of a specific Superfund site reuse, in contrast to the negative values we report. Of course, there can be many site-specific differences, but a key difference between the two studies is our inclusion of the (very large) external

---

[6] The curious relationship of the model's net present value and the discount rate is due to the negative NPV of the results. If NPV is negative, an increase in the discount rate increases net present values.

costs and our reluctance to include additional difficult-to-measure benefits from remediation, e.g., recovered values from surrounding properties. Removing the former and including the latter would, of course, lead to increased net present values for all remediation-reuse scenarios.

Key Drivers and Uncertainties

The model's robustness was examined using sensitivity analysis, as exemplified by the tornado chart presented in Figure 5.7. The goal of sensitivity analysis is to determine how the model's results change with small changes in various parameter values. A model is *robust* if large changes in model parameterization do not affect the ultimate outcome; a model is *sensitive* if small changes in parameters result in large changes in model results. A model's key drivers are those parameters that significantly impact its results and interpretation.

Panel II of Figure 5.7 identifies the input variables having the greatest impact on a remediation-reuse scenario's overall project's net present value. Note that the rankings are remarkably similar across treatment regimes and reuse alternatives. Probability of survival, excess cancer risk, and the value of a human life are the three most important inputs across practically all scenarios. Unfortunately, these variables are among the most difficult ones to estimate, due to a lack of both scientific knowledge and social consensus.

The discount rate is also an important input variable, primarily as a result of its relationship with another important driver, *time*. Because we are evaluating the discounted sum of the expected future benefits and costs, timing has a significant impact on net present values. For one thing, some reuse alternatives may not be able to be implemented until remediation is complete. If the remediation activities take a long time, the reuse benefits will be pushed farther into the future, and consequently worth less (in present value terms). For example, at Emmell's Septic Landfill, we entertained two alternatives: ecological and residential reuse. We assumed ecological reuse of the site (as a pine barren) could begin while the site was being remediated, allowing for near-immediate gains from the provision of ecosystem services. On the other hand, redevelopment of the site into residential housing lots could not begin until the site was fully remediated and it was safe for people to actually live on the property.

Timing and the discount rate also affect the magnitude of the present values of the external costs by allowing for greater or lesser exposures to harmful substances. The faster a site is cleaned up, the smaller the present value of a given set of external costs. On the other hand, potentially responsible parties may actually *prefer* delays, because they decrease discounted cleanup costs (Rausser, Simon & Zhao 1998). This dichotomy highlights the importance of using an integrated and adaptable model such as the one we have described. Even though Treatment #4 is the most expensive of all the remediation schemes, it requires the least amount of time for cleanup to occur, resulting in substantially smaller external health- and productivity-related costs. (It also has the smallest standard deviation of NPV because it avoids uncertainties associated with the external cost figures in later years.)

More could certainly be said about the difficulties inherent in estimating the parameters of our model. With 118 variables related to various aspects of human health, ecology, engineering, toxicology, and finance, it's likely almost everyone would find some aspect or value with which to disagree. We found it interesting that the various net present values were significantly less sensitive to the costs of treatment alternatives than to judgments about certain social values (e.g., the value of a human life and appropriate discount rate).

Viscusi, Hamilton, and Dockins (1997) note that using mean values for risk assessment (as we have done in our model) may overstate cleanup costs because significant differences may exist with regard to exposure. On the other hand, it may not always be worthwhile to gather better information, as Fleming and Adams (1997) discovered in their work on environmental tax strategies to reduce nitrate pollution. Because we believe the chief value of our model is presenting a structure for evaluating alternative remediation-reuse scenarios against one another, as opposed to establishing an absolute valuation, we're less concerned about these challenges than others might be. As long as all alternatives are on the same playing field, so to speak, we're in good shape.

Stigma

A shortcoming of our model is its failure to account for the depressed values of surrounding properties due to the presence of a Superfund site. McCluskey and Rausser (2003) note that it's possible for an area to suffer from both temporary and long-term stigma. The former, an "environmental externality," goes away when the site is cleaned up, but the latter, a "neighborhood externality," may persist for many years. One extension of our work would be to incorporate the expected benefits of reduced environmental stigma using data from hedonic pricing studies, similar to Kiel and Zabel's (2001) work on the Woburn site.

Implications

The proximate implications of our work with the SARR model are that in at least one case, redevelopment of a Superfund site for ecological reuse provides higher net benefits than redevelopment as residential housing, implying that greater attention should be paid to the measurement and incorporation of ecosystem services in decision-making. Moreover, it is important to consider both means and ends together. While ecological reuse dominates residential reuse for any given treatment alternative, the situation may change if a particular treatment alternative is selected for some nonfinancial reason. For example, the NPVs of the Treatment #1-Ecological Reuse scenario and Treatment #4-Residential Reuse scenario are quite similar. On a strictly financial basis, the two are nearly equivalent, suggesting that the ultimate decision will probably be based on so-called extrafinancial issues. Our work also contributes to the growing literature on the analysis of site-specific Superfund remediations.

Ultimately, we hope our approach to the modeling of Superfund site remediation alternatives will see greater use as a more general structural modeling tool, and that it will inspire additional discussion on such key issues as the valuation of natural capital, probabilities of survival, excess cancer risks, the value of a human life, discount rates, uncertainty analysis, and the determination of exposure levels.

CONCLUSION

The SARR model described in this chapter provides a means for organizing and analyzing inputs associated with redevelopment decisions for Superfund sites under an adaptive management approach. By explicitly including values for ecosystem services and allowing for different treatment regimes and end-uses, it provides decision-makers with an objective, yet flexible technique that incorporates both costs and benefits for multiple reuse scenarios, including ecological reuse.

The model can be used to demonstrate that when key factors affecting reuse of a Superfund site are considered before the remediation phase, decisions can be better tailored to support cost-effective reuse of the site. The SARR model argues that better decisions can be made by transparent consideration of external costs, and by explicitly taking into account benefits flowing from a site's ultimate reuse objective.

Our model directly supports the adaptive management goals presented in the introduction to this chapter. It provides site managers and other interested parties with a framework allowing the joint consideration of decisions concerning remediation and reuse alternatives. The model can be easily reparameterized as new or better information becomes available, and is even flexible enough to accommodate different remediation strategies and reuse alternatives if they were to arise as well. In other words, managers need not reinvent the wheel each time they wish to consider different alternatives or evaluate their current management scheme.

Adaptive management recognizes that our environment and society are in constant change. As a result, decisions made today might not be ideal tomorrow. Therefore, by employing a management strategy that inherently considers the dynamic nature of the environment and society, we can make changes that will improve the overall management of our lands. Models such as ours can help decision-makers to consider different alternatives and to reevaluate their choices, ultimately leading to improved Superfund site management decisions.

REFERENCES

Arrow, K. R., R. Solow, P. Portney, E. E. Learner, R. Radner, and H. Schuman. 1993. "Report of the NOAA Panel on Contingent Valuation." *Federal Register* 58 (10): 4602-4614.

Bateman, I. J., and K. G. Willis, eds. 2001. *Valuing Environmental Preferences: Theory and Practice of the Contingent Valuation Method in the US, EU and Developing*

Bonnani, S. J. 2002. Personal communication.

Cannon, Jonathan Z. 2006. "Adaptive Management in Superfund: Thinking Like a Contaminated Site." In *Reclaiming the Land*, Gregg P. Macey and Jonathan Z. Cannon, eds., 47-85. Amsterdam: Springer.

Champ, P. A., T. C. Brown, and K. J. Boyle, eds. 2004. *A Primer on Nonmarket Valuation.* Dordrecht: Kluwer Academic Publishers.

Chichilnisky, G., and G. Heal. 1998. "Economic Returns from the Biosphere." *Nature* 391 (12 February): 629-630.

Clemen, R. T., and T. Reilly. 2001. *Making Hard Decisions with Decision Tools.* Pacific Grove, CA: Duxbury Thomson Learning.

Costanza, R. 2000. "Social Goals and the Valuation of Ecosystem Services." *Ecosystems* 3:4-10.

Costanza, R., R. d'Arge, R. de Groot, S. Farber, M. Grasso, and B. Hannon. 1997. "The Value of the World's Ecosystem Services and Natural Capital." *Nature* 387:253-260.

Cummings, R. G., D. S. Brookshire, and W. D. Schulze. 1986. *Valuing Environmental Goods: An Assessment of the Contingent Valuation Method.* Totowa, NJ: Rowman and Allanheld.

Daily, G. C., ed. 1997. *Nature's Services: Societal Dependence on Natural Ecosystems.* Washington, DC: Island Press.

de Groot, R. S., M. A. Wilson, and R. M. J. Boumans. 2002. "A Typology for the Classification, Description and Valuation of Ecosystem Functions, Goods and Services." *Ecological Economics* 41:393-408.

Dolman, A. J., E. J. Moors, and J. A. Elbers. 2002. "The Carbon Uptake of a Mid-latitude Pine Forest Growing on Sandy Soil." *Agricultural and Forest Meteorology* 111 (3): 157-170.

E^2 Inc. 2002. *Superfund Benefits Analysis.* Charlottesville, VA: E^2 Inc.

El Serafy, S. 1998. "Pricing the Invaluable: The Value of the World's Ecosystem Services and Natural Capital." *Ecological Economics* 25:25-27.

Environment Canada. 2005. *Environmental Valuation Reference Inventory.* http://www.evri.ca/english.

Farber, S. 1998. "Undesirable Facilities and Property Values: A Summary of Empirical Studies." *Ecological Economics* 24:1-14.

Farrow, S., and M. Toman. 1999. "Using Benefit-Cost Analysis to Improve Environmental Regulations." *Environment* 41 (2): 12-21.

Fleming, R. A., and R. M. Adams. 1997. "The Importance of Site-specific Information in the Design of Policies to Control Pollution." *Journal of Environmental Economics and Management* 33 (3): 347-358.

Forman, R. T. T., ed. 1979. *Pine Barrens: Ecosystem and Landscape*. New York: Academic Press.

Fountain, J. C. 1997. "Removal of Nonaqueous Phase Liquids Using Surfactants." In *Subsurface Restoration,* C. H. Ward, J. A. Cherry, and M. R. Scalf, eds., 199-207. Chelsea, Michigan: Ann Arbor Press, Inc.

Geschwind, S. A., J. A. J. Stolwijk, M. Bracken, E. Fitzgerald, A. Stark, and C. Olsen. 1992. "Risk of Congenital-Malformations Associated with Proximity to Hazardous-Waste Sites." *American Journal of Epidemiology* 135 (11): 1197-1207.

Goodwin, B. K., L. A. Offenbach, T. T. Cable, and P. S. Cook. 1993. "Discrete-Continuous Contingent Valuation of Private Hunting Access in Kansas." *Journal of Environmental Management* 39 (1): 1-12.

Goulder, L. H., and R. N. Stavins. 2002. "Discounting - An Eye on the Future." *Nature* 419 (6908): 673-674.

Gund Institute for Ecological Economics. 2005. *The Ecovalue Project*. University of Vermont. http://ecovalue.uvm.edu/evp.

Hamilton, J. T., and W. K. Viscusi. 1999. "How Costly is 'Clean'? An Analysis of the Benefits and Costs of Superfund Site Remediations." *Journal of Analysis and Management* 18 (1): 2-27.

Heal, G. 2000. "Valuing Ecosystem Services." *Ecosystems* 3:24-30.

Herendeen, R. A. 1998. *Ecological Numeracy: Quantitative Analysis of Environmental Issues*. New York: John Wiley & Sons.

Huang, C. H., and G. D. Konrad. 2001. "The Cost of Sequestering Carbon on Private Forest Lands." *Forest Policy and Economics* 2 (2): 133-142.

Ibbotson Associates. 2005. *Stocks, Bonds, Bills and Inflation 2005 Yearbook*. Chicago: Ibbotson Associates.

Johnson, B. L., and C. T. De Rosa. 1997. "The Toxicologic Hazard of Superfund Hazardous Waste Sites." *Reviews on Environmental Health* 12 (4): 235-251.

Johnson, B. L., and C. T. De Rosa. 1999. "Public Health Implications." *Environmental Research* 80:S246-S248.

Jones, D. W., K. S. Redus, and D. J. Bjornstad. 2000. "The Consequences of Alternative Management Goals: A Non-linear Programming Analysis of Nuclear Weapons Legacy Clean-up at Oak Ridge National Laboratory." *Environmental Modeling and Assessment* 5:1-17.

Keeney, R. L., and H. Raiffa. 1993. *Decisions with Multiple Objectives: Preferences and Value Trade-offs*. Cambridge: Cambridge University Press.

Kiel, K., and J. Zabel. 2001. "Estimating the Economic Benefits of Cleaning Up Superfund Sites: The Case of Woburn, Massachusetts." *Journal of Real Estate Finance and Economics* 22 (2/3): 163-184.

Lange, D., and S. McNeil. 2004. "Clean It and They Will Come? Defining Successful Brownfield Development." *Journal of Urban Planning and Development* 130: 101-108.

Lybarger, J. A., R. Lee, D. P. Vogt, R. M. Perhac, Jr., R. F. Spengler, and D. R. Brown. 1998. "Medical Costs and Lost Productivity from Health Conditions at Volatile Organic Compound-Contaminated Superfund Sites." *Environmental Research* 79: 9-19.

McCluskey, J. J., and G. C. Rausser. 2003. "Stigmatized Asset Value: Is It Temporary or Long-term?" *Review of Economics and Statistics* 85 (2): 276-285.

Means Company, R. S. *See* R. S. Means Company.

Merkhofer, M. W., R. Conway, and R. G. Anderson. 1997. "Multiattribute Utility Analysis as a Framework for Public Participation in Siting a Hazardous Waste Management Facility." *Environmental Management* 21 (6): 831-839.

Meyer, P. B., and T. S. Lyons. 2000. "Lessons from Private Sector Brownfield Redevelopers: Planning Public Support for Urban Regeneration." *Journal of the American Planning Association* 66 (1): 46-57.

Millennium Ecosystem Assessment. 2005. *Ecosystems and Human Well-Being: Synthesis.* Washington, DC: Island Press.

Morgan, M. G., and M. Henrion. 1990. "Analytica: A Software Tool for Uncertainty Analysis and Model Communication." In *Uncertainty: A Guide to Dealing with Uncertainty,* 257-286. Cambridge: Cambridge University Press.

National Research Council. 1999. *Innovations in Ground Water and Soil Cleanup: From Concept to Commercialization.* Washington, DC: National Academy Press.

National Research Council. 2005. *Valuing Ecosystem Services.* Washington, DC: National Academy of Sciences.

Neill, P. 2002. Personal communication.

Newell, R. G., and R. N. Stavins. 2000. "Climate Change and Forest Sinks: Factors Affecting the Costs of Carbon Sequestration." *Journal of Environmental Economics and Management* 40 (3): 211-235.

Pearce, D. W. 1998. "Auditing the Earth: The Value of the World's Ecosystem Services and Natural Capital." *Environment* 40 (2): 23-27.

Pearce, D. W., and R. K. Turner. 1990. *Economics of Natural Resources and the Environment.* Baltimore: Johns Hopkins University Press.

Probst, K. N., D. Fullerton, R. E. Litan, and P. R. Portney. 1995. *Footing the Bill for Superfund Cleanups: Who Pays and How?* Washington, DC: Brookings Institution and Resources for the Future.

Probst, K. N., D. M. Konisky, R. Hersh, M. B. Batz, and K. D. Walker. 2001. *Superfund's Future: What Will It Cost?* Washington, DC: Resources for the Future.

Probst, K. N., and D. Sherman. 2004. *Success for Superfund: A New Approach for Keeping Score.* Washington, DC: Resources for the Future.

R. S. Means Company. 2002a. *Environmental Remediation Cost Data - Assemblies* (8th ed.). Kingston, MA: R. S. Means Company.

R. S. Means Company. 2002b. *Environmental Remediation Cost Data - Unit Price* (8th ed.). Kingston, MA: R. S. Means Company.

Rabl, A. 1996. "Discounting of Long-term Costs: What Would Future Generations Prefer Us to Do?" *Ecological Economics* 17 (3): 137-145.

Rausser, G. C., L. K. Simon, and J. H. Zhao. 1998. "Information Asymmetries, Uncertainties, and Cleanup Delays at Superfund Sites." *Journal of Environmental Economics and Management* 35 (1): 48-68.

Sagoff, M. 1997. "Can We Put a Price on Nature's Services?" *Philosophy and Public Policy* 17 (3): 13-17.

Salzman, J., B. H. Thompson, Jr., and G. C. Daily. 2001. "Protecting Ecosystem Services: Science, Economics, and Law." *Stanford Environmental Law Journal* 20:309-332.

Schneider, K. 1993. "EPA's Superfund at 13: Stains on the White Hat." *New York Times,* 6 September, 7.

Smedes, H. W., N. Spycher, and R. L. Allen. 1993. "Case History of One of the Few Successful Superfund Remediation Sites: A Site at Salinas, California, USA." *Engineering Geology* 34 (3/4): 189-203.

Sosniak, W. A., W. E. Kaye, and T. M. Gomez. 1994. "Data Linkage to Explore the Risk of Low-Birth-Weight Associated with Maternal Proximity to Hazardous-Waste Sites from the National-Priorities List." *Archives of Environmental Health* 49 (4): 251-255.

Toland, R. J., J. M. Kloeber, Jr., and J. A. Jackson. 1998. "A Comparative Analysis of Hazardous Waste Remediation Alternatives." *Interfaces* 28 (5): 70-85.

Toman, M. 1998. "Why Not to Calculate the Value of the World's Ecosystem Services and Natural Capital." *Ecological Economics* 25:57-60.

USEPA. 1994. *Common Cleanup Methods at Superfund Sites* (No. EPA 540/R-94/043). U.S. Environmental Protection Agency.

USEPA. 1999. *NPL Site Narrative for Emmell's Septic Landfill.* http://www.epa.gov/superfund/sites/npl/nar1549.htm (accessed September 19, 2005).

USEPA. 2003. *Reusing Superfund Sites.* http://www.epa.gov/superfund/programs/recycle/pdfs/reusingsites.pdf (accessed September 19, 2005).

USEPA. 2004. *Superfund: Building on the Past, Looking to the Future.* http://www.epa.gov/superfund/action/120day/pdfs/study/120daystudy.pdf (accessed September 19, 2005).

USEPA. 2005. *Superfund Budget History.* http://www.epa.gov/superfund/action/process/budgethistory.htm (accessed September 19, 2005).

Viscusi, W. K., and J. T. Hamilton. 1999. "Are Regulators Rational? Evidence from Hazardous Waste Cleanup Decisions." *American Economic Review* 89:1010-1027.

Viscusi, W. K., J. T. Hamilton, and P. C. Dockins. 1997. "Conservative versus Mean Risk Assessments: Implications for Superfund Policies." *Journal of Environmental Economics and Management* 34 (3): 187-206.

Willis, K. G. 1989. "Option Value and Nonuser Benefits of Wildlife Conservation." *Journal of Rural Studies* 5 (3): 245-256.

6

INSTITUTIONAL CONTROLS AT BROWNFIELDS:
Real Estate and Land Use, Not Just RCRA and Superfund

Jennifer L. Hernandez
Holland & Knight

Peter W. Landreth
Holland & Knight

INTRODUCTION

Communities in almost every state are affected by brownfield sites, from small rural towns to dense urban areas. Left untouched, brownfields pose environmental, legal, and financial burdens on communities and their taxpayers. When cleaned up, these sites can become powerful engines for economic growth and community vitality. Creativity and innovation at the local level promote collaboration between all levels of government, businesses, and nonprofit organizations to transform brownfields into economically productive and environmentally beneficial sites.

Brownfields are sites where expansion, redevelopment, or reuse of the site is complicated by the presence or potential presence of a hazardous substance, pollutant, or contaminant. (See Section 211(a) of the Brownfields Amendments, 42 U.S.C. § 9601(39)(A).) Various sources have estimated that there are somewhere between 400,000 and 1,000,000 brownfield sites in the United States (Smith 2001; Reid 2001; GAO 1995: 3; NADORF 1999: 4; EPA 2001). A diverse group of stakeholders, including federal, state, and local regulators, potentially responsible parties, developers, lenders, and the community, all have a strong interest in bringing these sites back into productive use. It is unrealistic and uneconomic in most cases to clean up brownfields to an unrestricted use standard; rather, by focusing on the intended reuse, the site can be cleaned up to a level determined by the environmental risks posed by the proposed use. This risk-based remediation is more cost-effective, but it results in residual contamination to properties, requiring the ongoing control of potential risks. Improving stakeholders' confidence in the long-term reliability and enforceability of institutional controls[1] designed to control these risks is critical to accomplishing this goal.

"Institutional controls" is used in this chapter to mean "legal or physical restrictions or limitations on the use of, or access to, a site or facility to eliminate or minimize potential exposures to chemicals of concern or to prevent activities that could interfere with the effectiveness of a response action" (*ASTM Guide*). Institutional controls are also referred to as "Activity and Use Limitations," or AULs; "Land Use Controls," or LUCs; or "Environmental Covenants" or "Servitudes." The U.S. Environmental Protection Agency (EPA) uses a narrower definition of institutional controls, namely, "non-engineered instruments such as administrative and/or legal controls that minimize the potential for human exposure to contamination by limiting land or resource use" (EPA 2000: 2).

CERCLA and RCRA have historically been focused on the cleanup of contaminated sites as the ultimate goal, with the potential to redevelop and reuse the site as a subsequent benefit. While the remediation of a brownfield site to environmental protectiveness standards is certainly the essential prerequisite for redevelopment, it is market forces, local real estate and land use realities, and investor dollars that ultimately drive the redevelopment. EPA has recognized the importance of considering "reasonably anticipated future land use" as well as requiring "community involvement" in selecting remedies at Superfund sites (Laws 1995: 3; EPA 2002b). However, this recognition needs to extend to the reality that market forces will not only inform the ultimate reuse of the site but will determine it.

Proceeding from this recognition, it becomes essential that local land use realities are incorporated into the remediation process as early as possible. The traditional model of first cleaning up a site and *then* determining how to reuse it is simply untenable from both (1) a real estate investor's perspective, due to the inefficiency and cost of having to potentially move dirt more than

---

[1] "Institutional controls" is used in this chapter to mean "legal or physical restrictions or limitations on the use of, or access to, a site or facility to eliminate or minimize potential exposures to chemicals of concern or to prevent activities that could interfere with the effectiveness of a response action." ASTM Standard Guide for Use of Activity and Use Limitations, Including Engineering and Institutional Controls, E 2091-00 (hereinafter, "ASTM Standard Guide" or "ASTM E 2091-00"). Institutional controls are also referred to as Activity and Use Limitations, or AULs; Land Use Controls, or LUCs; or Environmental Covenants or Servitudes. The U.S. Environmental Protection Agency ("EPA") uses a more narrow definition of institutional controls, namely, "non-engineered instruments such as administrative and/or legal controls that minimize the potential for human exposure to contamination by limiting land or resource use." EPA's "Institutional Controls: A Site Manager's Guide to Identifying, Evaluating and Selecting Institutional Controls at Superfund and RCRA Corrective Action Cleanups," p. 2 (Sept. 2000).

once, not to mention running the risk of having a remediated but unentitled property; and (2) from the local community's perspective, where consensus needs to be built based on not just the benefit of cleanup, but on the big picture benefits of the reuse to follow cleanup. In order for these sites to be efficiently and cost-effectively remediated and repurposed, the remediation process needs to be integrated with the entitlement and construction process: once the earth-movers begin to implement the remedy, they should be prepared and authorized to continue until the proposed reuse is completed. The earlier the ultimate intended reuse of the site can be determined, the more efficient and cost-effective the risk-based remedial selection can be.

The long-term management of the site can be simplified and streamlined by crafting these remedies prospectively to reflect the intended reuses of the site as well as the risks that have to be addressed under the applicable environmental regulatory requirements. With application of a broader view of the site that takes into account both the necessary short-term remediation and the long-term site prospects, these sites should be able to attract investment and maximize the vested interest of the local community. This approach requires the use of optimal institutional controls that (1) will ensure that risk-based environmental health and safety standards are met; (2) will be enforceable; and (3) will provide prospective investors with assurances as to future environmental liability.

These principles essentially dovetail with the adaptive management approach to Superfund sites discussed elsewhere in this book. In the Superfund context, adaptive management principles recognize that flexibility is required to accommodate future reuses of the site while complying with CERCLA's remediation requirements. This flexibility can be maximized by incorporating existing land use tools to put in place the kind of response mechanisms that can allow for adaptation years down the road of remediation. Effective institutional controls fulfill adaptive management principles by providing such mechanisms, meeting both agency and investor goals: remediate and manage site contamination and convert the site into profitable reuse. Institutional controls are developed with response to future changing conditions and contingencies in mind – exactly the kinds of events EPA is not well-equipped to manage in timely ways. Effective institutional controls can take site management off of EPA's hands, while still retaining EPA oversight, and put it in the hands of the vested local community.

Improving stakeholders' confidence in the reliability and enforceability of institutional controls is critical to the long-term success of efforts to bring brownfield sites back into productive reuse. For many, the conversion of brownfields into benefits presents an ominous challenge. Prospective purchasers of environmentally contaminated sites face the specter of liability under the Comprehensive Environmental Response, Compensation, and Liability Act ("CERCLA" or "Superfund") unless they can be confident of a defense to CERCLA's strict, joint, and several liability scheme (42 U.S.C. § 9607). Sections 221, 222, and 223 of the Small Business Liability Relief and Brownfields Revitalization Act of 2001 (the "Brownfields Amendments"), which amended CERCLA, provide parties with such a defense if they can establish that they are "in compliance with any land use restrictions established or relied on in connection with the response action at a vessel or facility" and that they did not "impede the effectiveness or integrity of any institutional control employed at the vessel or facility in connection with a response action." However, stakeholders must be confident in the efficacy, reliability, and enforceability of these institutional controls in order to ensure the successful remediation of brownfields.

Because of the many challenges facing federal and state regulators and other stakeholders, valuable tools that may aid in brownfields redevelopment may be overlooked. Viable institutional controls raise issues relating to both real estate and environmental law, and the proper implementation of institutional controls requires the cooperation and understanding of both real estate and environmental practitioners. The world of land use, which occurs mostly at the local level, provides a host of existing tools that may be the most effective institutional controls. Moreover, while many remedies are available to address the environmental and human health risks present at brownfield sites, the appropriate remedy must be selected with an eye to the ultimate use planned for the site, the nature of the risks posed by the site, and the local land use context. Land use compatibility is critical to the long-term effectiveness of the remedy.

Many of the sites where these principles will apply most palpably are those smaller-scale brownfield sites that may slip through the regulatory cracks because they are not hazardous enough to merit federal Superfund attention, but dirty enough to be encumbered with CERCLA liability. Recognizing that the stigma attached to such liability is usually enough to scare off potential investors, the brownfields movement promoted more localized state voluntary cleanup and brownfields programs ("VCPs"). These programs aim to establish clearer cleanup requirements, provide some limitations on liability, create financial incentives for redevelopment, streamline the governmental review process, and provide clear documentation when sufficient cleanup has been conducted. The utilization of institutional controls, incorporating risk-based cleanup principles, is a critical component of these programs. The effective implementation of institutional controls that are customized to the site-specific risks, the background land use context, and the planned redevelopment uses, will provide the certainty for investors who require assurances that they will not be saddled with crippling environmental liabilities.

The importance of effective implementation of institutional controls applies with equal force to Superfund and RCRA sites. CERCLA refers to the use of "enforceable measures" (e.g., institutional controls) as a potential component of a selected remedial alternative (see 42 USC § 9621(d)(2)(B)(ii)(III); *Site Manager's Guide* 2000: 3). EPA has recognized the need for flexibility in the cleanup and redevelopment of Superfund sites and has increasingly encouraged its site managers to implement institutional controls that can facilitate more cost-effective risk-based management. In its guidance for Superfund and RCRA site managers, EPA states that institutional controls "are vital elements of response alternatives because they simultaneously influence and supplement the physical component of the remedy to be implemented" (*Site Manager's Guide* 2). EPA notes (in the transmittal letter of the *Site Manager's Guide*) that while its institutional controls guidance is intended for Superfund and RCRA site managers, the same concepts should also apply to states, local agencies, and private individuals who are implementing institutional controls in the broader brownfields context.

EPA has described four basic categories of such institutional controls: proprietary controls, state and local government controls, statutory enforcement controls, and informational device controls (see, e.g., *Site Manager's Guide* 3-4). Proprietary controls include property use restrictions based on private property laws, such as restrictive covenants and easements. State and local government controls include local laws or permits, including zoning codes and building permits. Enforcement tools include documents that require individuals or companies to conduct or prohibit specific actions (e.g., consent decrees, unilateral orders, or permits). Informational devices include deed notices or public advisories that alert and educate a community's citizens about a brownfield site.

IMPLEMENTING INSTITUTIONAL CONTROLS

EPA has increasingly recognized the value of using institutional controls to complement engineering controls at cleanup sites. For example, a recent report by the General Accounting Office noted that the proportion of Superfund sites with institutional controls in place rose from 10% for Superfund sites deleted in the period from 1991 to 1993, to 53% for sites deleted in the period from 2001 to 2003 (GAO 2005: 10). Even more striking, 83% of the Superfund and 65% of the RCRA remedy decision documents finalized during the 2001-2003 period call for some use of institutional controls. However, despite the increased recognition and utilization of institutional controls in brownfield cleanups, long-term effectiveness remains a problem in the face of ongoing monitoring burdens, uncertainty related to the enforceability of certain institutional controls under varying state laws, and high transaction costs (GAO 2005: 10).

These problems underscore the need to take into account the local land use context at brownfields and to implement institutional controls based on existing tools that are consistent with that context. Cleanups often focus on engineering controls to control remaining contamination, such as the installation of a cap, slurry wall, groundwater treatment system, or other similar mechanism designed to physically control contamination or limit site access. However, engineering controls may be costly and require ongoing maintenance and inspection. Legal and administrative institutional controls can offer more cost-effective solutions.

All four types of institutional controls have their strengths and weaknesses, but the efficacy of institutional controls can be maximized by recognizing the applicability of existing real estate and land use tools. Of the four types of institutional controls, proprietary and governmental controls offer the most opportunities to capitalize on these existing tools and utilize the local legal background. But all four types of institutional controls, in conjunction with engineering controls, can and should be coordinated to provide the most cost-effective, long-term solution for brownfield sites.

The successful utilization of proprietary and governmental controls, such as imposing deed restrictions and adapting zoning codes, demonstrates that brownfields law, whether in the Superfund realm or the VCP context, is as much a creature of real estate and land use law as it is of environmental law. By recognizing this principle, practitioners can ensure that institutional controls make sense from both an environmental perspective – ensuring that the chosen remedy is protective of human health and the environment – and a land use perspective – ensuring that the remedy and the institutional controls implemented to enforce it utilize and complement the local land use context.

Proprietary Controls

Proprietary controls are based on the common law in each state and involve traditional property law. They involve legal instruments placed in the chain of title of the site or property. The most common examples of proprietary controls are easements and covenants. Methods for creating and enforcing proprietary controls will vary depending on the property's jurisdiction as well as on factors such as (1) whether the party implementing the proprietary control is a responsible party implementing the remedy, a facility owner/operator, or some other party; (2) whether the property owner is willing to retain or convey the necessary real property interests or enter into a covenant or other similar legal instrument; and (3) who the holder of the proprietary control will be (EPA 2002a: 9). These controls raise unique enforcement issues because third parties (including federal and state environmental protection agencies) typically do not have a direct right to enforce these controls.

Proprietary controls are commonly referred to as "deed restrictions," though this is not a legal term of art (*Site Manager's Guide* 2). Deed restrictions such as easements and covenants allow a landowner or developer to place certain restrictions or prohibitions on future uses and activities at the site. Collectively, different types of deed restrictions may (1) be used to encourage or mandate standards within a development, such as building sites, building and landscaping materials, and architectural guidelines; (2) limit the density of buildings, dictate the types of structures that can be erected or prevent buildings from being used for specific purposes or from being used at all; or (3) preserve open space, wildlife habitat, unspoiled views, and solitude. These are all features for which homebuyers are willing to pay a premium, and the adaptability of deed restrictions is readily applicable to brownfields.

Moreover, these types of controls fulfill the principles of adaptive management. Deed restrictions are traditional land use tools with which local communities and governments are familiar and comfortable. They provide built-in incentives for local communities to monitor and enforce the controls in place at brownfield sites on a real-time basis, providing much more rapid response to unexpected contingencies than EPA could offer under its five-year review process (see 42 U.S.C. § 9621(c)).

Deed restrictions may include traditional real property interests such as restrictive covenants, equitable servitudes, and easements. For example, an easement might permit the holder to access the land for monitoring; a restrictive covenant could be used to prohibit digging or disturbing the soil on a portion of the property. The creation of each of these instruments requires certain legal formalities, such as a writing, an intention by the original parties to place the restriction upon the land, vertical and horizontal privity of estate, and the requirement that the restriction "touch and concern" the land. Deed restrictions can be powerful, cost-effective tools for brownfields, as they can be implemented without the involvement of any federal or state authority and can be privately enforced. Since they run with the land, these instruments can provide long-term protection for sites with contaminants left in place. However, these instruments also raise difficult enforcement questions since they have traditionally only been enforceable by the instrument holder, and not by third parties, including federal and state environmental protection agencies. Moreover, such instruments are creatures of highly variable state laws, which may further limit who can hold and enforce a deed restriction.

With these limitations in mind, certain states and entities have taken steps to facilitate the implementation and enforcement of proprietary controls by governmental agencies or third party organizations. For example, easements may be conveyed to non-profit organizations such as community stewardship organizations or land trusts that have the long-term capacity to hold and monitor deed restrictions. Colorado passed legislation in 2001 allowing its Department of Public Health and Environment to hold and enforce "environmental covenants," which are binding upon current and future owners of the property (Colorado Revised Statutes §§ 25-15-101 *et seq.*). California has several statutory authorities enabling its Department of Toxic Substances Control ("DTSC") to enter into covenants that run with the land (Cal. Health & Safety Code § 25355.5(a)(1)(c)). DTSC is also authorized to require a property owner to record covenants imposing institutional controls as a condition of a permit (Cal. Health and Safety Code § 25202.5). A California property owner may also enter into a covenant agreeing to refrain from certain acts on his land, and thereby bind future owners, if the instrument containing the covenant is labeled "Environmental Restriction" and recorded. About half of the states have laws, varying widely in scope, that provide for land use restrictions to be used in conjunction with risk-based brownfields remedies, but the majority apply only to state-led cleanups (NCCUSL *Uniform Environmental Covenants Act* ("UECA"), prefatory note 6).

The National Conference of Commissioners on Uniform State Laws ("NCCUSL") has undertaken a more comprehensive national effort to draft a model environmental covenant that could be adopted by all 50 states and would apply to both state- and federal-led cleanups. This legislation would provide clear rules for state agencies to create, enforce, and modify an "environmental covenant," defined as "a servitude arising under an environmental response project that imposes activity and use limitations," to restrict the use of contaminated real estate (NCCUSL, UECA § 2.4). The Uniform Environmental Covenants Act would eliminate traditional common law requirements, such as vertical and horizontal privity and the "touch and concern" requirement, which can limit the efficacy of restrictive covenants. The NCCUSL notes two principal policies served by confirming the validity of environmental covenants: (1) to ensure that land use restrictions and other controls designed to limit the potential environmental risks of residual contamination will be reflected in land records and effectively enforced over time as valid real property interests; and (2) to return previously contaminated property to "the stream of commerce" (NCCUSL 1). The Act implicitly recognizes that redeveloping brownfields requires the combination of both environmental and real estate and land use principles.

Governmental Controls

The second category of institutional controls is governmental controls (state and local), including zoning and variances, building codes, drilling permit requirements, and groundwater regulations (*ASTM Guide* § 6.3; *Site Manager's Guide* 12-15). Governmental controls may be enforced by a governmental agency in a court action when there has been a violation of the control or when the agency can establish that there is an "imminent and substantial endangerment." For example, CERCLA § 106 allows an enforcement action for "imminent and substantial endangerment to public health, welfare, or the environment"; RCRA allows for corrective action for unpermitted facilities with "interim status" (42 U.S.C. § 6928(h)) or "imminent and substantial endangerment" (42 U.S.C. § 6973). As with the proprietary controls discussed above, governmental controls can be tailored to suit the specific site, and they maximize local resources. In addition to serving as useful institutional controls, they also increase the involvement of local governments, which EPA describes as serving three important roles: (1) determining future land use; (2) helping engage the public and assisting in public involvement activities; and (3) implementation and long-term monitoring and enforcement of institutional controls (*Site Manager's Guide* 7). Land use practitioners are already familiar with zoning codes and land use plans and the agencies that implement them, and incorporating these existing tools into the brownfields context can reduce the transaction costs associated with parcel-specific proprietary and engineering controls.

Local governments have broad authority to impose land use controls, building codes, and other restrictions that can be incorporated into brownfields remedies. At the highest level, local governments can use their general plans, as well as adopt narrower specific plans and redevelopment plans, to accommodate brownfields sites and their reuse. Layered on these general sources, effective zoning can be used to prohibit specific activities or types of development. Moreover, local governments and communities are familiar with enforcing these requirements and can respond more quickly than more detached and distant federal and state agencies.

Zoning can play an important role in structuring an effective remedy. In the dynamic world of zoning and development, land uses may be transient. Today's factory may be tomorrow's park. Zoning gives local authorities the ability to respond to new information or changing conditions in a fast and cost-effective manner. This is precisely the kind of tool that fits within an adaptive management framework at brownfield sites. If EPA identifies the ultimate reuse for the Superfund site early and involves the local government and local community to build consensus, local zoning controls can be used to increase the local community's vested interest, decrease EPA's workload, and provide for long-term, responsive site protection.

Zoning can complement parcel-specific proprietary controls by regulating land use, building height, area of structures, population density, and intensity of use. Zoning can be particularly useful as an institutional control since it can cover a large number of parcels that are affected by the chosen brownfield remedy. For example, a zone overlay could be used to prohibit residential development in an area with residual contamination, or restrict development in areas with contaminated groundwater.

In addition to these general land use controls, local governments can also use their police power to establish building codes, permit requirements, groundwater use restrictions, and other ordinances to control activities. Building codes can be used to regulate everything from the depth and height of buildings to the type of materials that may be used. By imposing building and grading permit requirements, local authorities can control the pace of development and maintain oversight of brownfields as they are redeveloped. Local governments can also restrict use of groundwater by establishing groundwater management zones or prohibiting well drilling. Other ordinances can prohibit access to certain areas or activities such as fishing or swimming in areas with residual contamination. Local governments can thus use their police power to implement a suite of institutional controls that can be locally monitored and enforced and, more importantly, more effectively incorporate brownfields and their ultimate reuse purposes into the overall local land use scheme.

However, governmental controls are not without their limitations. As EPA points out, "effectiveness of enforcement [of governmental controls] depends on the willingness and capability of the local governmental entity to monitor compliance and take enforcement action" (*Site Manager's Guide* 14). This concern is compounded when the site is a federally managed Superfund site, where EPA can enforce the implementation of institutional controls but not necessarily their long-term maintenance (*Site Manager's Guide* 2). That is, a local government may agree to change the zoning of a given Superfund site to prohibit residential use as part of the selected remedy, but EPA has no power to restrict the local government's authority to subsequently change the zoning designation in the future. This highlights the importance of incorporating local input into brownfields cleanups.

Nonetheless, the flexibility of land use controls can be a double-edged sword in the brownfields context. As mentioned above, local authorities can respond quickly to changing land uses by adapting their zoning controls. This can be problematic for stakeholders who depend on the long-term stability of institutional controls to safely contain residual contamination. To address this concern and to maximize the effectiveness of local governmental controls, it is important to anticipate future land uses early in the remediation process and to establish procedures to provide early notice of any changes in land use or zoning laws that may affect those assumptions.

Enforcement Controls

The third category of controls is statutory enforcement tools. Statutory enforcement controls include orders used by federal and state regulatory programs, consent decrees that may specify activities prohibited at a particular property, and permits that specify permitted and prohibited activities and uses on a property (*ASTM Guide* § 6.4; *Site Manager's Guide* 21-23). These instruments may be issued unilaterally or negotiated to compel a party to limit certain activities and to ensure the performance of affirmative obligations, such as monitoring and reporting requirements (EPA 2002a: 3). Controls such as consent decrees are commonly used at Superfund sites, but the long-term effectiveness of a consent decree may be undermined by the fact that, without the employment of other types of institutional controls, a consent decree is not by itself binding on subsequent property owners. Adaptive management principles are instructive on this point. If regulators integrate remediation and future reuse planning, otherwise static enforcement controls can be combined with other institutional controls to build flexibility and adaptability into remedies.

Informational Controls

The fourth category of controls is informational devices, such as the deed notice (*ASTM Guide* § 6.5; *Site Manager's Guide* 24-27). Informational devices are designed to ensure that, before concluding a real estate transaction, the parties are made aware of the environmental conditions on the property (e.g., chemical releases; restrictions on use, access, and development). Generally accepted types of notice that qualify as informational devices include record notice (in land records), direct or actual notice, and notice to a government authority, registry act requirements (requiring states to maintain a database of sites relying on institutional controls), and transfer act requirements (*ASTM Guide* § 6.5). Though common, informational controls may not prove particularly efficacious for the long-term management of a remediated site, since they do not actually limit or restrict the use of the property, nor do they provide a legal basis for regulators to prevent a property owner from disturbing or exposing remaining contaminants (GAO 2005: 32-33). However, informational controls serve an important purpose in providing transparency and facilitating monitoring and enforcement of other institutional controls.

MONITORING INSTITUTIONAL CONTROLS

Whatever institutional controls are implemented for brownfield sites, agencies charged with overseeing them, whether federal or state, face the long-term challenge of monitoring and enforcing those controls. Residual contamination may remain on-site until long after transactions have been completed and institutional controls established. The difficulties inherent in long-term monitoring underscore the value of using tools that rely on local, nearby enforcement, but local governments and organizations will also face challenges in monitoring brownfields. In its 2005 report, GAO found the ongoing monitoring and enforcement of institutional controls to be one of the most significant obstacles to their effectiveness (27). Describing the problems inherent in EPA's five-year review process under CERCLA, GAO cited one site that included an institutional control requiring EPA monitoring of worker safety precautions during digging – yet on GAO's visit digging was occurring unbeknownst to EPA, and the responsible EPA official had not visited the site since the previous five-year review, four years earlier (30).

New technology can help to address this problem. EPA has taken some steps in this direction, most notably by establishing the Institutional Controls Tracking System ("ICTS") in 2001, initially designed to track controls used at Superfund sites. In the RCRA context, EPA has modified RCRAInfo, its database of information on individual RCRA sites, to identify sites where institutional controls have been implemented. However, both of these resources are in their infant stages, and GAO notes that "information necessary to determine whether institutional controls are being monitored and enforced is not currently included" in either of these systems (2005: 41). Nonetheless, these resources are a start, and their development can facilitate the efficacy of institutional controls and brownfield cleanups. Also promising are private sector resources like Terradex, founded in 2002 to monitor institutional and engineered controls at private properties for both private parties and government agencies (see http://www.terradex.com). Providing long-term monitoring enforcement reliability can increase the stakeholder confidence that is essential to the success of institutional controls and to the larger goal of bringing brownfields back into productive use.

CONCLUSION

Superfund sites and brownfield sites in general must be viewed as more than simply dirty sites to be cleaned up. A contaminated site that is merely remediated and capped, without a concrete reuse in place, is not sustainable: there is no incentive for private redevelopment investment, and local community involvement and interest will be limited. A site, however, that holds the promise of profitable reuse, with a carefully thought-out plan combining remediation with the entitlement, construction, and financing processes, will revitalize a community and facilitate the goals of all brownfield stakeholders. This goal can be achieved by applying adaptive management principles, prospectively planning flexible institutional controls that will allow for the productive redevelopment of a site while controlling the risks of whatever remnant contamination there may be under the selected remedy. Brownfields should thus be structured against the backdrop of the local land use context, with the understanding that brownfields law, especially in the context of institutional controls, is an amalgam of environmental, real estate, and land use law. There is a wealth of tools available for structuring a viable remedy for contaminated land. Maximizing the use of existing land use and real estate tools at brownfields sites lowers transaction costs and leads to more effective enforcement. Where these controls are tied to direct supervision by the state environmental agency or EPA, contingencies will be difficult to manage responsively in the long term; but where these controls can be tied to local interests through zoning, deed restrictions, and other similar tools, enforcement can be accomplished largely at the local level, with EPA oversight streamlined to maximize thinly spread federal resources. Remedies work best when all the available tools of land use are employed in structuring the remedy: landowners assist in enforcing standards, and local authorities will conduct oversight in lieu of the federal government. The result is a more cost-effective and workable scheme for the management of Superfund and brownfield sites.

REFERENCES

American Society for Testing and Materials. 2006. *ASTM Standard Guide for Use of Activity and Use Limitations, Including Engineering and Institutional Controls.* E 2091-00. Cited as *ASTM Guide.*

ASTM Guide. See American Society for Testing and Materials.

EPA. *See* U.S. Environmental Protection Agency.

GAO. *See* U.S. General Accounting Office *and* U.S. Government Accountability Office (successive names).

Laws, Elliot P. 1995. Memorandum from Elliot P. Laws, Assistant Administrator, U.S. EPA, to Regional Directors. "Land Use in the CERCLA Remedy Selection Process." OSWER 9355.7-04. May 25, 1995.

NADORF. *See* National Association of Development Organizations Research Foundation.

National Association of Development Organizations Research Foundation. 1999. *Reclaiming Rural America's Brownfields.* December.

National Conference of Commissioners on Uniform State Laws (NCCUSL). 2003. *Uniform Environmental Covenants Act.* Washington, DC.

NCCUSL. *See* National Conference of Commissioners on Uniform State Laws.

Reid, U.S. Senator Harry. 2001. Statement. *Congressional Record* 147: S3904. April 25.

Site Manager's Guide. See U.S. Environmental Protection Agency. 2000.

Smith, U.S. Senator Gordon. 2001. Statement. *Congressional Record* 147: S3886. April 25.

U.S. Environmental Protection Agency. 2000. "Institutional Controls: A Site Manager's Guide to Identifying, Evaluating and Selecting Institutional Controls at Superfund and RCRA Corrective Action Cleanups." September. Cited as *Site Manager's Guide.*

U.S. Environmental Protection Agency. 2001. http://www.epa.gov/epahome/headline_011102.htm

U.S. Environmental Protection Agency. 2002a. "Draft Guide to Implementing, Monitoring and Enforcing Institutional Controls." December.

U.S. Environmental Protection Agency. 2002b. *Superfund Community Involvement Handbook.* EPA 540-K-01-003. 3rd rev. ed., April.

U.S. General Accounting Office (GAO). 1995. *Community Redevelopment: Reuse of Urban Industrial Sites.* June.

U.S. Government Accountability Office (GAO). 2005. *Hazardous Waste Sites: Improved Effectiveness of Controls at Sites Could Better Protect Public.* January.

7

RETHINKING COMMUNITY INVOLVEMENT FOR SUPERFUND SITE REUSE:
The Case for Consensus-Building in Adaptive Management

E. Franklin Dukes
Institute for Environmental Negotiation, University of Virginia

INTRODUCTION

Many citizens and communities are continuously buffeted by forces outside of their control that can dramatically impair their quality of life. The decisions of people with little or no stake in a particular neighborhood or community – agencies of state and federal government, developers of property, leaders of multi-national industries and banks – affect the choice and affordability of the housing in which community members live, the transportation they use, the food they put on their table, the way they communicate with the outside world, the means to educate their children, the work they pursue, and, especially, their health, through the quality of the air they breathe and the water they drink.

For those who care about this condition, the fundamental questions it provokes are profound: How can individuals and communities come together to make their own voices heard and shape their own neighborhoods and communities? How can those who help design those neighborhoods and communities understand their communities' needs and concerns? How can responsible public and private officials help give voice to the previously unheard and put a face on the previously unseen? How can communities overcome the actions of less responsible parties?

Many may be surprised to learn that valuable answers to these questions are being forged in locations where the most horrific cases of disempowerment and abuses of power have occurred, as communities determine new uses for Superfund sites. In this chapter I draw on lessons from those and other cases to present a way of thinking about community involvement for Superfund site reuse that reconceives community involvement as a long-term process of *community consensus-building* (Susskind, McKearnan & Thomas-Larmer 1999). This process shares many features of *adaptive management* (Holling 1978) and is part of a larger trend toward *collaborative governance* (Henton & Melville 2005; Walker and Senecah, in preparation; "A call to scholars" 2005) and what has been termed *adaptive governance* (Scholz & Stiftel 2005). This consensus-building process begins as early as possible after listing on the National Priorities List (NPL) and continues even as initial reuses are determined and implemented, rather than being a series of isolated, discrete processes that end upon construction completion.

The knowledge base for these lessons comes from many sources. These include, first, my own experiences as mediator and facilitator working across the United States with communities that have suffered the tragedies of heavily contaminated sites. That experience includes both brownfield and Superfund sites within communities seeking to determine and implement appropriate reuse. These sites include the following:

- A town in a Southern, low-income rural community attempting to reclaim an industrial site abandoned nearly a century earlier (Elliott and Bourne 2005);
- An affluent Pacific Northwest island struggling to recover from the economic loss of a creosote plant whose operations devastated the land and harbor, while at the same time honoring the memory of Japanese-Americans forced off the island and into internment at that very same site (http://wyckoffsuperfund.com/; www.bijac.org/memorial.html);
- Three jurisdictions in Northern Illinois seeking to bring recreational opportunities to the site of a contaminated landfill (www.epa.gov/superfund/programs/recycle/landfill.htm);
- A large gold mine on the eastern edge of the Rockies (www.epa.gov/region8/superfund/giltedge/index.html);
- A former chemical manufacturing facility located in the middle of a mid-sized town seeking to find economic opportunity (Wilkinson et al. 2003); and
- A heavily contaminated site in an industrialized river and a 330-acre adjoining parcel of land that includes large industries, small businesses, and residences, as well as a cherished community church.

One of the privileges of serving as a mediator or facilitator in situations of this sort is to be told stories by many people with stories to tell.

A second source of knowledge is a set of unpublished case studies put together by the faculty and students of the University of Virginia's Center for Expertise in Superfund Site Recycling (see http://www.virginia.edu/superfund/). The primary research was conducted by teams of student architects and planners, who investigated in depth a half-dozen sites preparing for reuse. These sites represent a cross-section of size (including one that involved a cluster of three sites within one geographic region), sources of contamination, type of original use (e.g., mining, orchard, landfill, naval base, manufacturing), potential for reuse, and other relevant factors. Assisted by at least one faculty member, each team interviewed key parties, dug through historical records, and examined factors such as market potential, physical characteristics, and remediation characteristics, while integrating considerations of culture and community into the tales of environmental and economic woes. These studies demonstrate that contaminated sites are more than properties with toxic chemicals embedded in the ground or water. They are real places, with real histories. Each place has embedded within it a community's dreams, culture, and history.

Another source has been the successes that have occurred throughout the country in bringing contaminated sites to legitimate, productive reuse. Better-known places such as the Avtex Fiber site in Front Royal, Virginia (Bromm & Lofton 2002), the Industri-Plex site in Woburn, Massachusetts (Wernstedt & Hersh 1998; Wernstedt & Probst 1997), and many others less well known, demonstrate that even highly complex sites with long, ugly, and sometimes tragic histories can find eventual uses that are supported by the community (see http://www.epa.gov/superfund/programs/recycle/success/index.htm for stories of reuse). Those successes embody sometimes heroic efforts made by citizens, site owners, and state and federal agency personnel. I have learned the stories of these successes both anecdotally at places such as the annual Superfund Redevelopment Partnership and Superfund Community Involvement conferences (www.epa.gov/superfund/action/community/ index.htm) and in more formal settings, including the stories told by participants in reuse planning efforts filmed by the EPA's Superfund Redevelopment Initiative (see www.clu-in.org/search/t.focus/id/423/). Although much of what follows is critical of unproductive community involvement practice, my criticism of such practice is based upon the knowledge that meaningful community involvement has been successful in leading to uses that are economically productive as well as legitimate in the eyes of formerly skeptical community members.

PROBLEMS WITH COMMUNITY INVOLVEMENT

A Failure of Imagination

Ask any group of public officials – state regulators, EPA staff, local elected office holders, community planners – whose responsibilities include controversial, severely contaminated sites about their experience with "community involvement," and you may well receive a pained expression followed by a story – or several – about the frustrations and indignities of working with communities. These stories follow similar patterns: unreasonably angry citizens who don't understand the realities of rules and regulations designed to protect the public; misunderstandings turned into suspicion, and even hostility, by people who know little about science and risk but a lot about raising a ruckus; good intentions and hard work frustrated by insufficient resources and attention.

Ask a group of citizens who live adjacent to such sites about their experiences, and you will often hear a similar refrain from at least some portions of those neighbors, albeit with different targets of complaint. The site owner(s) avoid real contact with the community, and maybe the truth; the EPA representatives, whose names and faces change periodically even as their terminology continues to bewilder, drop in and out of the community in seemingly random patterns of intense activity interrupted by long periods of silence; state and local officials appear uninformed, uninterested, or even willing accomplices to corporate mendacity. Not all sites, of course, have such histories. But every case that I have worked on has had some level of complaint of this sort. Several have had a substantial opposition group with a history of antagonism between public, private, and community members. Indeed, the brownfields community pilot facilitation project evaluated by Elliott and Bourne (2005), the Superfund community involvement program (see http://www.epa.gov/superfund/action/community/), and the Technical Outreach Services for Communities (TOSC) program (see http://www.toscprogram.org), which provides support to citizens to help them understand issues of contamination, are each responses to the failures engendered by intense community fears and conflict.

Both perspectives can be accurate. Citizens can behave badly – and why not, when faced with a legacy of broken trust and real threats to health and welfare? Community advocates may see any particular contaminated site as just one of many failings or even systematic and intentional mistreatment by government and business (Cole and Foster 2001). And community involvement may be conducted poorly. That is, at times – some citizens would say most times – what is already known to be effective at promoting meaningful involvement does not occur. Information is slow to be released, legitimate community concerns are discounted, plans are changed without warning, records are lost, and worse (Cole and Foster 2001). Middle- and upper-class, highly educated, overwhelmingly white business leaders and public officials may have little experience and less interest in finding ways to engage and satisfy community members whose lives are most impacted by the former's decisions but who often share none of those individual characteristics (Matsuda 1995).

A deep critique of community involvement was raised in a seminal article by Arnstein (1969). Arnstein observed that community involvement can range from manipulation and therapy, which she terms non-participation; to tokens of information provision, consultation, and placation, with no intention of taking public views into account; to authentic partnership and actual citizen control. Her context was an era in which public and private authorities often assumed that their power over "have-not citizens, presently excluded from the political and economic processes..." was the natural order (216). That era may be over; at least that paradigm has been superseded by laws, regulations, and policies that promote an ideal of community involvement including not only an informed and engaged public but willing partners in agency staff and elected officials. But many problems associated with community involvement remain, particularly at contaminated sites. As Cole and Foster (2001: 16) observe, "Current environmental decision-making processes have not been effective in providing meaningful participation opportunities for those most burdened by environmental decisions."

Assessing the quality and effectiveness of community involvement is as challenging a task as assessing "government," or "the public" itself. Community involvement is used for many purposes and conducted in many ways. Community involvement can mean a posted notice in the Federal Register with a time-limited opportunity to make written comments. Or it can mean a public hearing or a series of such hearings. Or it may mean convening a multi-stakeholder, consensus-seeking group, along with any combination of such processes. Those public hearings may be poorly advertised and held in locations and at times that make it hard to attend, or they may be located where affected populations can most easily attend. The consensus-seeking group may be nothing more than the same old faces, or it may represent the faces and voices of the community.

Beierle and Cayford (2002, 2003) screened 1,800 case studies of public participation for environmental decisions and investigated 239 of those cases intensively. They note that 30 years ago, community involvement for environmental decisions consisted most often of fulfilling statutory obligations to hold public hearings and respond to public comment; such involvement was intended primarily to ensure at least a minimum of government accountability. The goals of community involvement continue to include such accountability; but now community involvement is expected to contribute to better decisions, to resolve conflict and build trust, and to improve capacity for future problem-solving. They identify five social goals of public participation for environmental decisions. These include:

- incorporating public values into decisions;
- increasing the substantive quality of decisions;
- resolving conflict among competing interests;

- building trust in institutions; and
- educating and informing the public (Beierle & Cayford 2003: 54).

They found a mixed record of success in meeting these goals through community involvement. Educating participants scored fairly high; building trust, fairly low. But they also found that community involvement tends to reach participants who do not represent the same socioeconomic characteristics of the public actually affected by such decisions, and educating the wider public most often does not occur.

Thus we see that despite a massive academic and popular literature (see http://www.coastal.crc.org.au/toolbox for an annotated bibliography of over 500 references related to science and community involvement), numerous comprehensive how-to manuals (see, for instance, http://www.epa.gov/publicinvolvement/involvework.htm for many "how-to" references and case studies), and significant investment within agencies in human resources, research, and training, intense dissatisfaction with the interaction of public officials and private citizens occurs on all sides. This status is disheartening but not surprising; the combination of high human health and economic stakes, complex issues, and incomplete information makes for a volatile recipe no matter how well-intentioned or skilled community members and those in authority may be.

Yet it is not these structural obstacles alone that leave a community disengaged, upset, or fractured, and consequently unable to achieve viable reuse. Rather, this inability reflects something more fundamental: it is, in fact, the result of a failure of imagination and of faith (Lederach 2005). That is, too few can envision a way to bring diverse sets of people together in the face of fear, conflict, and uncertainty, to create what is yet unseen and unimagined by most of those people.

A senior public health advocate removed by a few weeks from a mediated session with tobacco farm leaders once advised me in frustration, "There *is* no common ground" (Dukes 2004). She was correct in one, limited sense: in conflict, common ground is rarely found merely by looking for it. But she was incorrect in a more important sense: like the farmland that produces tobacco's golden leaf, common ground needs to be cultivated in order to flourish. Cultivation means a willingness to take risks, commitment to the search for common ground, multiple skills, and a good bit of faith to envision a fruitful harvest when all that is immediately apparent is barren (or, to push the metaphor into reality, contaminated) dirt.

CHALLENGING STANDARD ASSUMPTIONS OF COMMUNITY INVOLVEMENT

Conventional thinking has its place. But several fundamental truths that apply to contaminated sites and the communities within which they are found contravene conventional thinking and the (typically implicit) assumptions that underlie standard community involvement procedures. These truths that contradict conventional thinking must be acknowledged and honored for effective, legitimate reuse to occur.

I suggest that, whether implicitly or explicitly, a number of assumptions underlie conventional practice in contaminated site reuse. These assumptions as a whole may not reside within any policies or manuals, nor do they necessarily reflect malevolent intention or dereliction on the part of individuals who hold many of these assumptions. Rather, they reflect such structural realities as competing demands for legitimate involvement and speedy reuse, funding limitations, and personnel policies, as well as the extraordinary difficulty inherent in finding legitimate and productive uses for heavily contaminated sites with many and varied stakeholding interests. These assumptions follow:

1) Community involvement is an add-on to the real work of site assessment, remediation, and reconstruction.

2) Community involvement consists primarily of responsible agencies offering information and gathering feedback in order to make effective, legitimized decisions.

3) Community involvement takes place as a series of discrete processes that need only occur at certain times ordained by statute and regulation. Community needs, concerns, and interests in reuse will remain relatively consistent throughout any planning process.

4) Community safety and health concerns can intrude upon a reuse planning effort.

5) Community involvement can be conceived and actualized by following a set of rules, procedures, and technologies.

6) The goals of community involvement are limited to helping inform reuse decisions and avoiding conflict.

Some of these assumptions may sound plausible. In certain circumstances, such as those involving sites with little controversy, few citizen concerns, and little likelihood of any significant impact of reuse on other community interests, the standard approaches to community involvement that follow from these assumptions may be sufficient to allow for reuse. But few contaminated sites that have been added to the National Priorities List – indeed, few contaminated sites at all – fit that profile.

Wernstedt and Hersh's (1998) analysis of community involvement at Superfund sites demonstrates the challenges. The mosaic of stakeholders and often conflicting interests includes site owners and potentially responsible parties fearful of liability and exorbitant cleanup costs, municipalities seeking new tax revenues, public health officials and citizens concerned about past health damage and present and future health risks, elected officials (and others) concerned with protecting their power base, and community activists hoping to revitalize their community. Active reuse markets satisfied with containment rather than removal may confront equity issues, private mechanisms to escape liability may provoke confrontations with advocates for transparency in decisions, and the impacts of site remediation and reuse may extend well beyond any one municipal jurisdiction, making for a disconnected patchwork of meetings and other community involvement processes.

Williams et al. (2001) note substantial structural barriers to effective community involvement at weapons management and disposal sites, barriers that occur at other contaminated sites as well. They observe that there is no agreement about what constitutes the "public" and what community involvement should be, they question whether the public can participate meaningfully when technical issues are complex, and they argue that we have little knowledge about whether and how citizens actually want to be involved. Finally, they add that widespread public distrust of government adds yet another barrier to effective involvement.

Conventional thinking is insufficient to address those barriers. What follows is my challenge to that conventional thinking, the assumptions that underlie it, and the barriers that it helps to perpetuate.

1) *Meaningful community involvement is – and must be treated as – an essential component of remediation and reuse.* Many contaminated sites have created "contaminated communities" with associated fear and distrust. Rebuilding public trust is not only desirable; in many cases it is a *necessary* condition to produce legitimized uses of the site. Community involvement, in short, matters.

2) *Reuse requires a shift beyond conventional community involvement toward community consensus-building.* This shift is necessary as decision-making power about reuse is diffused among several parties, and traditional

roles and authority give way to less hierarchical (and hence potentially more chaotic) decision structures.

3) *Community needs, concerns, and plans change as the community learns and interest and knowledge about reuse grow. Hence efforts to build consensus for reuse will require a long-term, adaptive effort.* It may take years from when a site is first place upon the NPL to construction completion for remediation of a site, and the reuse planning process may need to continue long after that. Planning cannot be static; indeed, an adaptive approach to reuse planning will help as knowledge about the site increases and community understanding grows. The community *can* learn, and that learning needs to be nourished and actively sustained beyond the boundaries prescribed by formal community involvement.

4) *Health and safety concerns are never an intrusion upon a reuse planning process.* Community concerns about health and safety cannot be sidestepped without eroding legitimacy and provoking community opposition.

5) *Community consensus requires above all a* commitment *to community and stakeholder empowerment; consensus can be helped by, but never actualized through, reliance upon community involvement techniques and technology.* Information sharing and effective meeting knowledge and skills are necessary, but insufficient; to build authentic consensus with a knowledgeable and engaged community requires intention and commitment above all.

6) *Consensus-building for reuse offers an opportunity for rebuilding community and renewing civic engagement.* Just as contaminated sites can be restored and made productive, contaminated communities can be made whole.

I will consider each of these principles in turn.

RECONCEIVING COMMUNITY INVOLVEMENT

1. Meaningful community involvement is – and must be treated as – an essential component of remediation and reuse.

The tragedy of contaminated communities

Leo Tolstoy in *Anna Karenina* wrote that "All happy families resemble one another; each unhappy family is unhappy in its own way." Similarly, each community with a contaminated site has its own dynamics. These dynamics often represent the elements of an authentic tragedy. The first act of this tragedy, which might be titled "Everything is Fine," finds early voices of concern drowned out by assurances of safety and economic demand. The script runs something like this:

Act One: **Everything is Fine**. In this Act, there is no problem or the problem has already been addressed.

Act Two: **We Know What We're Doing**. During this time, the voices of concern become noticeable. There may be an admission of some past harm but the problems are past, and, besides, people need to remember all the jobs provided to the community.

Act Three: **We Are Working On It**. By ACT III, the voices of opposition have gained strength. Nearby residents may be desperately fearful of harm from contamination, from job losses, or both. Any admission of harm may also be accompanied by assurances that clean-up will occur.

Act Four: **Whose Problem Is This?** By ACT IV, the tragedy is fully realized, the plant or landfill is crippled or closed, and jobs are gone. And so, after what may perhaps be literally years' worth of efforts to get someone to listen, and perhaps a few more years waiting to hear confirmation of the nature and extent of any harm, while enduring fears about health and economic effects, and undoubtedly many years of uncertainty about what will happen to the site as well as blame for the site owners, public officials, neighbors who complained, and neighbors who refused to complain, we arrive at:

Act Five: **Reuse – or, Everything Is Fine**. At this point, the chorus of voices asserting that "everything will now be fine, trust us because we know what we are doing, and think of all the new jobs," may well have lost its power. Remember that many community members have seen the first Act and are anticipating, or rather fearing, a sequel not unlike the original.

Overlay this context, which is present in various forms at many contaminated sites, with the sentiment that community involvement is subservient to more significant issues, and you will find what occurs with many such sites: highly antagonistic citizens who will fight public officials and private business tooth and nail for control over their own destinies. Some authorities realize the importance of community involvement, particularly those who have seen the power of achievement unleashed by effective community work at the aforementioned sites in Woburn, Massachusetts, or Front Royal, Virginia. Yet it is my experience that even relatively enlightened public officials who sincerely want community input nonetheless consider community involvement a distraction from "real" work – site assessment, remedial actions, development plans, and new construction. Eisen (1999) points out that public participation on contaminated sites has been seen as a threat to timely redevelopment with public officials fearful of lengthy delays and withdrawal of potential developers; indeed, he notes that many state brownfields statutes limited opportunities for public input for this very reason.

> At one meeting involving a community-based stakeholder group that I observed, an official tasked with engaging the community described, with some apparent satisfaction, that he had spoken with a librarian a few years earlier about potentially storing site-related materials, and that he had contacted the high school at that time about the possibility of hosting a public meeting at some point. He had done this shortly after the site made the National Priorities List (NPL), some five years earlier. This had been the extent of community engagement.

> Five years after the listing, the Remedial Investigation/Feasibility Study (RI/FS) still awaited approval, and concerned citizens wondered what was happening. This light flurry of activity – not even a flurry, but rather a dusting, and all that was claimed by way of community involvement during that period – was held up as an example by another official of their demonstrated commitment to community involvement. This claim did not impress community members.

Yet that attitude must change, and indeed is changing. Eisen's assessment of successful brownfields reuse reveals the emergence of a significant change in attitude, as participants report that community involvement is key to that success. Pepper's review (1997) of 20 case studies emphatically reports that almost all of the successful projects included early and significant community involvement. Elliot and Bourne (2005) detail how enhanced community involvement can transform sites and communities. And certainly EPA itself has many resources that promote meaningful involvement (see, e.g., U. S. Environmental Protection Agency 2001).

2. Reuse requires a shift beyond standard community involvement toward "stakeholder and community consensus-building."

Despite the beginnings of this shift, the standard conception of community involvement still dominates practice and writing. This standard model of community involvement involves an agency offering information and gathering feedback in order to make effective, legitimized decisions (Beierle & Cayford 2002; Arnstein 1969). Tools tend to be limited to distribution of informational material through handouts and web sites and a limited number of public meetings, where public and private officials share news and respond to questions from the public. This type of community involvement has its place, but that place is not a locality with a severely contaminated site awaiting reuse.

As Williams et al. (2001) confirm, government and industry often define community involvement in ways "diametrically opposed" to what is wanted by the public. While citizen groups want "active participation in decision-making, government and industry instead often provide public meetings and hearings" (44). Williams et al. (2001) declare that providing an opportunity for involvement is insufficient: successful community involvement generates an environment "in which the public wants to become involved" (46).

For Superfund sites, the typical conception of community involvement for questions of listing, assessment, and even remediation presumes EPA as decision-maker and the community as interested public (Teske 2000; U.S. Environmental Protection Agency 2004). But for redevelopment, those roles are altered substantially. Decision-making authority about land uses comes largely from the community rather than EPA, with "community" being defined as local interests: the site owner(s), elected officials, private businesses and developers, neighbors, and others with interests in land use.

Furthermore, successful reuse of contaminated sites, or at least the substantial proportion of such sites that may generate controversy, requires not just information sharing but negotiation, agreement, and ultimately cooperation among many diverse interests and institutions. Stakeholders may include: the U.S. EPA; state, tribal, and local governments; neighborhood groups; non-governmental organizations; and various parties from the private sector, including of course the site owner as well as prospective purchasers. Each of these interests may offer competing visions for the site and, depending upon the balance of power, such competition may (and often does) evolve into conflict and stalemate. Even when overt conflict is absent, the sheer number of parties and diversity of interests require coordinated planning.

The numbers and types of stakeholders, the scope of activities, the opportunities for conflict over competing visions for the site, and the legacy of contamination (including not just stigma but possibly active fear and resistance) mean that redevelopment of Superfund sites presents significantly different challenges from those that occur during listing, assessment, and remediation stages. I thus see a need to shift away from thinking about community involvement as an authority exchanging information and ideas with an interested public (and associated impressions as described by Arnstein), to a conception of community involvement as *stakeholder and community consensus-building.*

Consensus-building as a form of environmental conflict resolution has a history dating back over three decades, with a substantial body of theory and practice (see, e.g., Susskind, McKearnan & Thomas-Larmer 1999) and research (see, e.g., Dukes 2004). Characteristics of consensus-building include the following (Dukes 2004: 92):

- direct, face-to-face discussions;
- deliberation intended to enhance participants' mutual education and understanding;
- inclusion of multiple sectors representing diverse and often conflicting perspectives;
- openness and flexibility of process; and
- consensus or some variation other than unilateral decision-making as the basis for agreements.

The consensus-building process may or may not include a mediator or facilitator.

This shift moves beyond what English et al. (1993) term "stakeholder involvement." Stakeholder involvement, in their sense, consists of the use of representative groups, such as advisory groups and task forces, who offer advice but still do not share decision authority, and may operate by traditional majority-minority vote. Ashford and Rest (1999) warn that "over-reliance" on such elite stakeholder processes may take away from more community-focused processes, which then risks disempowering those who might be most closely affected by reuse decisions.

A shift to consensus-building enhances community processes; it combines traditional outreach (e.g., public meetings, mailings, television, radio, and print news) with stakeholder and community decision groups and their powerful tools of dialogue, deliberation, and multi-party decision-making. Consensus-building of this sort also suggests a host of first- and second-order benefits, including community trust, new partnerships, cumulative knowledge, and social and political capital (Innes 1999) that only engagement of this sort can provide.

Most significantly, this shift acknowledges the community as an authentic partner in decisions that may have significant effects on that community. Such a shift actualizes what is often promised but not offered: "meaningful" community involvement. Cole and Foster (2001: 16) describe this shift best:

> "Meaningful," in this context, means substantive dialogue among administrators, experts, and affected communities along with the opportunity for affected communities to influence the decision-making process. This means early, direct, and collaborative public participation. More important, it presupposes a power-sharing process in which government is but one party to the ultimate decision or agreement.

Building consensus also builds legitimacy, which is essential for reuse (Williams et al. 2001). Many localities, businesses, or other organizations have found it difficult to win support for reuse with key stakeholders and community members. There can be many reasons for opposition. Some of these may be appropriate, such as when a proposed use is too costly, or creates a public nuisance. Other reasons simply get in the way of beneficial reuse. Some of the most significant obstacles include:

- *Past site history.* Mistrust and fears of additional contamination and human health threats may dominate a community's response to potential reuses of a contaminated site.
- *Lack of awareness.* Simple misinformation about what has occurred at a site and what might occur can stall reuse plans.
- *Lack of consensus-building commitment and skills.* Many people and groups simply ignore the need to build consensus with a broad array of stakeholders for a site's use. They think instead that while "decide, announce and defend" (Cole & Foster 2001: 110) might be discredited elsewhere, it will work at their site.

What follows are some of the fundamental elements of a consensus-building process with multiple parties as derived from my experience. As the section to follow will show, the extent to which each element can be used effectively by community leaders, agencies, site owners, and others who might be leading a reuse planning effort depends to varying degrees on whether they are pursued as part of an adaptive management approach to the reuse of a contaminated site.

1. *Offer access to information* - When information is limited or slow to be revealed, the assumption of stakeholders is not that leaders know best, but that they are hiding something they don't want others to know. Constraints on information create confusion, contribute to uncertainty, and exacerbate any previous suspicion and antagonism that individuals or groups might have about the site or authorities. Leadership that becomes known for prompt

responses to requests for information and for providing access to its records and meetings will have a reserve of legitimacy when it makes mistakes or deals with sensitive issues.

It is particularly important to clarify how decisions get made: who is doing what, how issues will be addressed, what opportunities for involvement exist, and when decisions will be made and implemented.

2. Involve others immediately and offer adequate time for meaningful engagement - Too often, groups wishing to avoid controversy delay opportunities for involving other stakeholder or public interests until those affected by proposed plans have no recourse other than protest or disengagement. Yes, decisions often need to be made quickly. But early deadlines and short time limits can heighten uncertainty and increase suspicion about the motivations of parties responsible for those time constraints. Such suspicion is heightened when an issue is controversial or the group has a reputation for imposing unreasonable time pressures.

When deadlines are necessary they should be announced and publicized as far in advance as possible, and at any rate as soon as they are decided. The reasons for those deadlines should be made clear. Concerns about timing should not be dismissed out of hand, and when possible they should be addressed in ways responsive to the concerned parties.

3. Begin with needs and encourage options - Leaders often will propose a plan of action or a limited number of options in order to get discussion started. But interested parties presented with a single solution or a limited range of options often assume that their input is not wanted or valued. By beginning with a statement of need and encouraging the generation of options, issues are less likely to become polarized into "yes/no, either/or" situations.

This is particularly true for contaminated sites, where many parties may still have concerns about the condition of the site and ongoing health and safety risks. These individuals may well become supporters of reuse, but need to have their concerns heard, understood, and addressed before they can focus on reuse planning.

4. Be inclusive - Exclusion of individuals or groups from decision-making processes not only gives an impression of secrecy and fosters a sense of victimization, it wastes the potential ideas and support that excluded parties might bring. And inclusiveness means more than merely having open meetings; it may mean targeting special groups for mailings or other forms of publicity, bringing meetings to where interested parties are, and creative efforts to encourage involvement despite the many inevitable obstacles.

5. Focus on issues and respect the dignity of all parties - The standard response to criticism is to characterize it as obstructionist and blame it on the personal failings of those voicing concern. All too often the next step down the path to an all-out conflict is to demonize the opposition.

One can acknowledge the meaning and importance of an issue to individuals or groups who voice concern without agreeing with their answers to those concerns. Seeking to clarify and understand the sources of concern may well uncover a basis for new learning and agreement (see Fisher & Ury 1981). At worst, treating both critics and supporters with courtesy will model and encourage appropriate behavior for both groups. At best, it will earn respect and trust for you and those who support your interests.

6. *Accept responsibility when you are wrong* - While common excuses for the weaknesses or failures of plans and programs may have elements of truth, they do not lead to improved performance. They also create a perception of weakness and immaturity. Accepting responsibility does not have to mean that you are to blame, but that you will do what is necessary to get the job done.

7. *Create a culture of openness, inclusion, creativity, and respect* - The best way to destroy personal and institutional legitimacy is to practice consistently the arts of secrecy, defensiveness, rationalization, deception, and animosity. The best, and only, way to develop trust is to be trustworthy. It sounds trite, but say what you will do, and then do what you say.

8. *Convene a representative, consensus-seeking decision group to spearhead the reuse effort* - The sites whose reuse plans have the broadest support from stakeholders and the general community have taken the time to seek out, understand, and work through all the issues and obstacles that can impede reuse. For many of those sites, this has meant convening a consensus-based decision-making group with broad representative participation.

The power of consensus lies in the process in which decisions are made. Consensus requires meeting the needs of each member of the group, a requirement that forces groups to seek creative solutions that might not occur otherwise. Consensus processes can moderate the majority/minority dynamic that develops in voting groups. Consensus processes not only change how groups make decisions; they change how groups approach problems.

Consensus processes can be powerful, but they are useful only to the extent that participants fully understand and value the requirements of the process. Not all situations are suitable for a consensus-based group process. Consensus decision processes require time, stamina, and sufficient resources to provide high-quality information, sufficient time to work through conflicts and to learn, and active participation and commitment of group members to the process (Innes 1999). Reasons for the use of consensus decision processes include:

- Participants who will have some responsibility for implementing agreements need a say in decisions;
- It is important to get all parties to the table, and individual participants who might be skeptical of working with opponents or

those they don't know are reassured by having effective veto power over any decisions;

- Group members know that they need to attempt to satisfy the needs of all participants;
- Minority views which may have been summarily dismissed need to be given real consideration;
- A norm of responsibility for the group may be enhanced; and
- As a practical matter, decisions with broad-based support are more likely to be implemented.

9. Offer different interests different opportunities for engagement in reuse planning – one shoe does not fit all - Consider three levels of interest and engagement at any contaminated site:

1) *Key decision-makers without whose agreement redevelopment cannot occur.* These include the property owner(s); local government and other entities with land-use authority; the EPA and state agencies that have to sign off on any remediation plan and possibly approve particular types of reuse.

2) *People and organizations directly affected by the redevelopment.* These may include immediate neighbors; organizations that might develop or locate on the site; organizations that might promote a particular use for the site (such as a school system seeking playing fields, or an organization seeking habitat preservation); and any other people or organizations without whose endorsement redevelopment may be more difficult or may not occur.

3) *The broader community that may be impacted by the reuse.* These would include community taxpayers, consumers, those who would participate as users of the site facilities, or neighborhoods not directly adjacent to the site but that might be affected by noise or traffic.

These three levels require different processes of engagement, depending on the dynamics of a particular site (English et al. 1993). Some sites may require little attention to these elements, as the key decision-makers are in agreement and the community is not affected or shows little interest. For other sites, the behavioral dynamics may be the single greatest determinant of successful (or unsuccessful) redevelopment. Some sites may require little or no involvement of levels two or three; others may require active engagement and direct participation in key decisions of people in level two and even level three.

> Following are ideas generated by a steering group co-facilitated by the author for generating interest in a reuse project at a contaminated site. The group had several goals: to inform, to seek advice and support, and to recruit more individuals in specific tasks related to the project
>
> - Develop a project newsletter that could be mailed to Town residents and other interested parties.
> - Develop a Town newsletter that would cover this project as well as other issues.
> - Conduct a "listening project" that would involve training a core group of citizens to go door-to-door to provide information about the project and then listen to citizen ideas and concerns.
> - Develop a project listserv that would both provide up-to-date information about the project and that would solicit views of citizens.
> - Develop contacts with churches and other civic organizations that could publicize project plans and activities.
> - Use banners and posters and other visible notices on the site and in other prominent places.
> - Conduct a survey of citizen awareness of, and ideas for, the project.

If a consensus-building group process is appropriate, balanced and inclusive representation is essential (Beierle & Cayford 2003; Cole & Foster 2001). Even in non-voting situations, such as might be found in consensus processes where any single member can veto any decision, the number and type of representatives makes a difference. That difference can be seen in the issues raised, the amount of time spent considering various issues, the weight given particular options, and other group dynamics, as well as the end result. Considerations include:

- Legitimacy - a group that is viewed as representative will have legitimacy that a group seen as excluding interests will not have, and decisions are less likely to be attacked;
- Equity - balanced and diverse representation is inherently fair, independent of any practical reasons attached to that representation;
- Diversity of interests and ideas - diversity of representation can bring broader knowledge and new ideas and innovation, even if such diversity sometimes makes group functions more difficult;
- Accountability - the ability to confer with, and report concerns of, a variety of organizations and constituencies is important for decision implementation; and
- Group dynamics - too many like-minded people can create barriers to effective decision-making, including an "us vs. them" attitude and insular thinking.

In the introduction to the study, "Citizens and Politics" (Kettering Foundation 1991), Kettering Foundation president David Matthews argues that citizens long to restore the integrity and vitality of public discussion. Citizens truly want officials and others to listen to their concerns; they yearn for open discussions both among themselves and with public officials. One study found that a significant proportion of neighbors want to participate in the redevelopment of contaminated and abandoned sites (Greenberg & Lewis 2000). That opportunity needs to be offered.

3. Community needs, concerns, and plans change as the community learns and interest and knowledge about reuse grow. Hence efforts to build consensus for reuse will require a long-term, adaptive effort.

Nobody would assume that a damaged stream could be returned to health by simple manipulation of one or two variables, or that restoration could be accomplished in a few short months. Yet for some reason, few anticipate that the human systems involved at and near Superfund sites – many as complicated as their ecological counterparts, and many if not most intricately intertwined within that ecological system – need as much attention and care, as much patience, and as much administration as the ecological.

Adaptive management is a relatively simple concept with profound consequences for reuse. First articulated by Holling (1978), adaptive management has evolved during the past decades into several types of management regimes. Fernandez-Gimenez (forthcoming) offers one of the clearest explanations of adaptive management and one that offers appropriate context for reuse consensus-building. She notes that people have been learning by doing and adding to that knowledge over generations; adaptive management is a term applied to a particular type of such learning, one in which scientific experimentation can be applied in a deliberate manner to natural resource management. The rationale for this has to do with the need to take action to manage natural systems despite their complexity and unpredictability. Learning can be accelerated and enhanced and management strategies altered when they do not meet desired goals. She distinguishes between *active* adaptive management, which tests a variety of different actions at the same time and may allow for manipulation of several variables, and *passive* adaptive management, which attempts to include the same design elements and learning goals as active adaptive management while testing only one action at a time.

Like ecological management, consensus-building for reuse at any one site requires continued evaluation and adjustment throughout the process. If we accept the parallel between natural and human systems (while acknowledging that the distinction is somewhat artificial, in that humans are part of natural systems), their common complexity and unpredictability means that no immutable formula for community involvement can provide an effective process blueprint.

Consider the many dynamic variables involved in planning for reuse:

- the extended and unpredictable timetable between listing and construction completion (according to the EPA, this can last up to several decades (http://epa.custhelp.com);
- the tendency for agency personnel to change (one site I worked on had gone through three different EPA Remedial Project Managers even before the Remedial Investigation/Feasibility Study was well under way, and others changed community involvement coordinators);
- the initial uncertainty about the nature and extent of contamination, which can continue into the remediation process;
- the inability at times to predict exactly how effective remediation strategies may be until such strategies are implemented;
- dynamic political environments, including multiple political jurisdictions;
- changing markets;
- evolving and sometimes conflicting community preferences.

Furthermore, while this is most obvious on larger sites (e.g., of mineral extraction), all such sites are themselves parts of ecosystems and include the same types of environmental dynamics that demand adaptive management in the first place.

At some level, people may recognize the scope of this human challenge. Indeed, I suspect that it is precisely the fear that the human dynamics are too complex, too unpredictable, and too messy that may lead site owners, public officials, and citizens alike to fear authentic engagement with one another over the issues of reuse. This is the same kind of phenomenon that can cause individuals who suspect they have some untreatable health problem to avoid consulting a physician: "Perhaps if I pretend hard enough, my problem will disappear."

Thus, planning for future reuse must be an iterative process of testing activity, reflection, and evaluation of outcomes, with decisions to proceed based on an assessment of likely outcomes. It cannot follow any formula or guidance that does not allow for this adaptive approach.

"Future reuse" may actually be misleading; some sites may have continued uses of portions of the property throughout the process of investigation, listing, RI/FS, and remediation. At the Central Chemical site in Hagerstown, Maryland (Wilkinson et al. 2003), reuse planning occurred even as renters conducted several types of business (not all of them legal, if rumors were accurate). Remediation may take place in phases, with portions of a site available for use before others are ready. Planning has to be continuous and flexible to allow for developments that occur during a period that may last as long as a decade or more. An adaptive approach offers an additional benefit of acknowledging that contingencies such as changing stakeholders, newly elected officials, discovery of additional contamination, availability of funding, and real estate market fluctuations are not interruptions, but instead normal dynamics that can and should be anticipated.

The time to begin to consider eventual site reuse is immediately upon listing of the site. The earlier that reuse becomes a consideration, the more likely that reuse planning may result in use or uses that reflect stakeholder and community interests. In fact, the EPA is required to consider the "reasonably anticipated future land use" (RAFLU) when determining a remedy (OSWER Directive No. 9355.7-04, Elliott Law, Assistant Administrator, May 25, 1995), although the amount that RAFLU factors into such decisions remains unknown.

One significant reason to begin discussing reuse early has to do with community perceptions. For citizens first confronted with the realities of living near a Superfund site, the notion that the site may eventually be reused safely may not make any sense. Involvement by citizens at problem sites typically is motivated by concerns about negative community impact (Williams et al. 2001), not by interest in reuse. An authentic consensus-building effort likely will need a long time for such concerns to be addressed, and ensuring that health and safety concerns are considered at the same time as reuse allows appropriate linkage between the two.

No planning process can guarantee that eventual reuse will occur in a way exactly consistent with its findings. However, a recommendation for reuse that is produced by an open, inclusive, informative process is more likely to have legitimacy (Leach & Sabatier 2005) and a certain momentum toward implementation than one produced behind closed doors or by a closed circle of participants.

The planning process, however, does not stop upon determination of any site use plans. Without continued advocacy, the likelihood of implementation may decrease as time passes, government and other organizations change personnel, and economic conditions fluctuate.

The best way to hold everyone accountable for commitments is to prepare beforehand to make these commitments genuine and feasible. At the end of a long planning process, people may be reluctant to discuss what happens if someone can't or won't live up to their promise. But the preparation made up-front can pay dividends in the long run. Specific ideas for ensuring accountability include:

- Put agreements in writing and make agreement language clear and explicit;
- Think through "what if" contingencies before they happen and emotions get high;
- Consider third-party support (from a trusted organization not involved directly in the issues) for monitoring and evaluating decisions;
- Consider building into the agreement a clear process for revision and modification; and
- Plan follow-up sessions whose agenda includes an accounting of how decisions have been implemented and a process for revising them if they have not had intended results.

Leadership of reuse planning efforts typically comes from local government, which has authority to regulate land use. However, some state or federal facilities or even privately owned sites may convene a consensus-building effort as well. The sponsor/convenor, whether a public or private body, should have public legitimacy and should maintain a link to decision-makers. If the sponsor/convenor is a private body, extra efforts to secure public legitimacy may be necessary to develop support for the project among community participants.

4. Community health and safety concerns are never an intrusion upon a reuse planning process.

Most people have heard or read about disasters that occurred after authorities proclaimed that "everything is safe" or "the risk is low." They may even have heard those assurances before contamination on a specific site was verified or risk was determined, only to learn later that the contamination or risk was real. If such concerns exist, they need to be acknowledged and incorporated into the reuse planning process.

> At one site, the original plan developed by the City called for the citizen advisory group to have three sub-committees: one to work on reuse assessment, by developing site reuse alternatives; one for public outreach, education, and participation, responsible for organizing public meetings in coordination with City planning staff; and one to seek resources and funding.
>
> During the first two meetings of the citizen advisory group it was apparent that concerns about health and safety had not been addressed through other processes. In fact, some members suspected that serious health problems they and family members suffered might have been caused by operations at the site.
>
> Despite concerns from some staff that this would derail the reuse planning process, a safety sub-group was established to identify and research concerns. The Technical Outreach Services for Communities (TOSC) program provided technical expertise at no cost to help address the community's site health and safety concerns. Although results were inconclusive, this legitimate effort satisfied concerned citizens that their needs were being considered. A consensus-based reuse plan was developed with the support of those who were most concerned about safety along with the other community members (Wilkinson et al., 2003).

5. Community consensus requires above all a *commitment* to community and stakeholder empowerment; consensus can be helped by, but never actualized through, reliance upon community involvement techniques and technology.

Recall the elements of tragedy described earlier. *All* contaminated sites represent unfulfilled promises. For some contaminated sites, community members have endured years of repeated denials, refusals, cover-ups, delays, and outright scorn from individuals in positions of authority. In some communities, these individuals used their authority as shields against scrutiny and, sometimes, against truth – shields of science or engineering, regulatory policy and enforcement, insider knowledge, or financial gain. Whether this was done intentionally or not, public trust was degraded (Williams et al. 2001).

Rebuilding trust requires demonstrating commitment to behave in ways that deserve trust (Leach & Sabatier 2005; Leach & Pelkey 2001). Only with a commitment to understanding all the public's fears, concerns, needs, and aspirations, and an equivalent commitment to addressing them honestly and to the greatest extent possible, can the process techniques offered in manuals and trainings be put to effective use.

6. Consensus-building for reuse offers an opportunity for rebuilding community and renewing civic engagement.

This last reconception of the role of community involvement will perhaps be the most challenging for many to accept. The vast literature about contaminated sites and their reuse describes remediation, regulatory compliance, financing, legal and regulatory reform, institutional controls, and a myriad of scientific, economic, political, and legal issues. Some of the literature takes community involvement seriously. But nowhere can be found any linkage of contaminated sites with "building community" and "civic renewal." One might even anticipate that for some public officials, site owners, and potentially responsible parties, successful reuse may be defined by mere survival of a sort; that is, if the headaches caused by agency demands, community fears, financial pressures, protests, and various other troubles that accompany NPL listing can just go away, or even be reduced to a level that is manageable, that's good enough.

That gauge is a poor substitute for what is possible if authentic consensus for reuse is sought. For while conflict may bring problems, it also brings opportunity.

> When I sat down with concerned citizens and public officials at a first meeting to discuss the possible reuse of an abandoned industrial site in a small town of 2,000, they spoke very little about the site. Their concerns were for the health and vitality of the community. In order for their reuse planning effort to be successful, they identified the following desired outcomes:
>
> - A stronger sense of community and civic-mindedness, with community members becoming engaged in the redevelopment project;
> - A community spirit that positive change is possible;
> - A stronger link with the county government;
> - Development of self-sustaining leadership to carry the project forward, as current leadership naturally evolves.
>
> Because they understood the fundamental importance of civic capacity building, they were successful at planning and implementing their reuse plans. Indeed, the work of these leaders now serves as a model for other communities (Elliott and Bourne 2005).

I have already observed that reuse may be a long-term process that involves many parties, who must be prepared to adapt to changing and often challenging circumstances. How can communities develop the leadership and the capacity to meet those circumstances? Innes (2002), examining how an active consensus-building organization builds capacity, highlights the research of Chaskin (2001). Chaskin observes that community capacity requires four elements: a sense of community, a level of commitment among members, an ability to solve problems and to translate commitment into action, and access to resources. High among the essential skills required for this capacity are abilities to work collaboratively with others and to resolve conflict.

A reuse effort that simply seeks to identify a preferred reuse may do just that in a short period of time; however, given the dynamic environment in which reuse planning occurs, what guarantees are there that any recommendations are legitimate, effective, and will be carried through to implementation? Unless sufficient long-term community capacity is developed to ensure that the recommendation is enacted, plans may lie on a shelf. Community capacity, Chaskin found, develops as members recognize their stake in the well-being of a place and are thus willing to act to support that locale. Consensus-building offers an opportunity to develop such interests.

The EPA supports such capacity-building in several ways, including the following (U.S. Environmental Protection Agency 2005):

- Awarding Technical Assistance Grants to 276 communities affected by Superfund cleanup efforts;
- Providing educational and technical support for more than 200 communities through the Technical Outreach Services to Communities (TOSC) program; and
- Organizing Community Advisory Groups in 90 communities across the nation.

But, as Teske (2000) argues, while technical assistance grants can support citizen participation, the degree of democratic governance to emerge depends greatly on the willingness of the legally empowered decision-making agency to allow citizen groups to influence the process. Research into collaborative groups (as well as common sense) indicates that the more involved individuals are in determining how a group operates, the more likely they are to care about and actively participate in the group (Innes 1999).

A burgeoning literature on deliberative democracy (see, e.g., Bourne 2002; Dukes 1996; Lukensmayer & Brigham 2005) recognizes the power of meaningful community engagement. Cole and Foster (2001: 112) observe that "the common good does not reflect merely aggregated and bargained preferences of the participants." When done effectively, consensus-building moves people beyond individual self-interest toward shared values: "In a deliberative process, citizens thus *create* the common good through discourse, as opposed to *discovering* it through preexisting preferences" (113).

My experience has been that when presented with authentic opportunities to engage officials and fellow citizens, many "ordinary" citizens are capable of responding with commitment, compassion, and integrity. As Barber (1984: 189) observes, talk has the power to make private self-interest into something "that makes possible civility and common political action." Talk nourishes empathy, and empathy develops bonds and promotes public thinking.

Residents of a contaminated community not only look for common ground, they seek *higher* ground (Dukes 2000). Such higher ground is provided by forums that both seek and reward principled behavior. The metaphor includes new ground, which means an opportunity to explore and discover that which is as yet unimagined; a new and enlarged perspective, which is a view not only of the whole picture, but also of how each individual fits within that picture; and a refuge, or forums for deliberation that provide a safe haven from the incivility and outright nastiness that too often accompany conflict. Reaching higher ground can be a challenge, but when achieved, the accomplishments can serve as beacons for others to follow.

With a commitment to authentic consensus-building, planning for reuse provides one opportunity to develop this higher ground.

CONCLUSION

Scholars are beginning to draw connections among environmental consensus-building, deliberative democracy, and adaptive management, with *adaptive governance* (Scholz & Stiftel 2005) and *collaborative governance* (Henton & Melville 2005) used to describe cumulative efforts that are inclusive, deliberative, and consensus-seeking to deal with recurring problems. This newly emerging paradigm is still in its infancy, and many obstacles make its acceptance problematical. Entities concerned with promoting reuse thus need to adopt policies supporting widespread and authentic deliberation (Forester 1999) as well as long-term community capacity building (Foster-Fishman et al. 2001).

Thus a final recommendation – not a challenge to conventional thinking, but important nonetheless – is the need to take advantage of the learning that occurs across many contaminated sites. True adaptive management involves a cumulative process of testing and experimentation across sites (Fernandez-Gimenez, forthcoming). Superfund site reuse represents an extraordinary laboratory for experimentation and innovation, and one whose lessons can extend to many other forms of community development and planning as well.

The analogy between adaptive management for ecosystems and consensus-building for reuse is by no means complete; but the comparison can be useful and there are real lessons from adaptive management applicable to the consensus-building effort. Figure 7.1 makes such a comparison. Most of the adaptive management statements are adapted from Lee's (1999) authoritative appraisal of adaptive management.

Adaptive management for conservation

The need for adaptive management is grounded in the acknowledgment that people do not know enough to manage ecosystems with predictable results.

Adaptive management focuses on managing the people who interact with the ecosystem.

Adaptive management is learning while doing.

Adaptive management explores questions to which there are few reliable answers by experimentation.

Adaptive management affects social arrangements and how people live their lives.

Adaptive management conducts experiments to build understanding about the ecosystem's processes and structures.

In adaptive management, learning requires active participation from those most likely to be affected by actions. Those who depend upon the resource may be those who know most about the condition of the ecosystem.

Conflict in adaptive management is inevitable and essential, and should be conducted in ways that the disputing parties perceive as legitimate, or it will thwart learning.

Adaptive management is likely to be costly and slow; this approach is reasonable for use in complex natural systems, including those most disturbed by human impacts.

Building authentic consensus for reuse

The need for consensus-building is grounded in the realization that no single party can know enough to determine what will be an appropriate and legitimate reuse.

Building authentic consensus requires a focus on managing how people engage one another to foster mutual understanding, empathy and problem-solving capacity.

Effective consensus-building requires a strong educational component to avoid falling back on the lowest common denominator for agreement.

Consensus-building does not seek an answer that already exists; rather it creates answers that can only be developed from the combined experience, resources, and ingenuity of diverse participants.

Authentic consensus requires understanding that site reuse affects how people live their lives.

Consensus-building needs to be given sufficient time to incorporate evaluation and methods for implementing change.

Consensus requires active participation from those most likely to be affected by reuse. Those who will be most directly impacted by reuse may know most about the community, economic, and even ecological dynamics that must be considered for reuse.

Conflict in consensus-building is inevitable and essential, and should be conducted in ways that the disputing parties perceive as legitimate, or it will thwart learning.

Consensus-building takes time and resources; it is most appropriate in situations that are complex and that involve significant community disturbance.

Figure 7.1. Lessons from adaptive management that can improve consensus-building for reuse.

What is required to build a storehouse of cumulative knowledge across sites can also be helpful in applying an adaptive approach to any particular site. Key elements include:

- Planning for a more enduring institutional memory in order to minimize the disruptions as various parties come and go. These changes come as a surprise but in fact should be anticipated. Disruptions are caused by everyday factors such as changes in political power, normal EPA and other agency personnel changes, and transient populations;
- Inculcating a serious, substantial, shared evaluation process that begins during the stage of project design, engages all parties in its design, implementation, and learning, and builds on past experiences (see Innes 1999);
- Ongoing evaluation of any consensus-building or stakeholder engagement processes, and incorporation of lessons learned into planning processes. This is what Yaffee (2003) terms "adaptive project management."
- Developing criteria for success that are site-specific but consistent with national program imperatives. Do not focus on the number of handouts produced and distributed, meetings held, and so forth as indicators of success. Those are inputs that may help achieve success but that cannot substitute for real outcomes such as increased knowledge about the site, perceived legitimacy of the reuse plan, improved relationships among parties, and commitments to implementation of reuse plans.

The EPA has taken many steps to disseminate learning about meaningful community engagement across sites. Guidance documents, case studies, videos, and an annual community involvement conference all are useful tools for knowledge development and exchange. A next step is to develop even more ways of sharing knowledge across sites through processes that engage researchers with program staff, local government, industry, and community members. The adaptive management framework is one in which learning is an ongoing and encouraged process; as contaminated sites and the communities in which they can be found evolve, such learning can help provide reuse options that are community-endorsed, safe, and legitimate.

REFERENCES

"A Call to Scholars from the Collaborative Democracy Network." 2005. *National Civic Review* 94 (3): 64-67.

Arnstein, S. 1969. "A Ladder of Citizen Participation." *Journal of the American Institute of Planners* 35:221.

Ashford, N. A., and K. M. Rest. 1999. *Public Participation in Contaminated Communities.* Cambridge: Massachusetts Institute of Technology, Technology and Law Program.

Barber, B. 1984. *Strong Democracy.* Berkeley and Los Angeles: University of California Press.

Beierle, T. C., and J. Cayford. 2002. *Democracy in Practice: Public Participation in Environmental Decisions.* Washington, DC: Resources for the Future.

Beierle, T. C., and J. Cayford. 2003. "Dispute Resolution as a Method of Public Participation." In *The Promise and Performance of Environmental Conflict Resolution,* R. O'Leary and L. B. Bingham, eds., 53-68. Washington, DC: Resources for the Future.

Bourne, G. 2002. "Democracy and Civic Engagement: To What Extent Do Consensus-Building Processes Improve Democratic Participation and Decision Making?" In *Critical Issues Papers,* S. Senecah, ed., 70-85. Washington, DC: Association for Conflict Resolution.

Bromm, S., and J. Lofton. 2002. "Negotiations in Superfund Cases – The Role of Communities in Site Redevelopment." Paper presented at the Sixth International Conference on Environmental Compliance and Enforcement, San Jose, Costa Rica.

Chaskin, R. 2001. "Defining Community Capacity: A Definitional Framework and Case Studies from a Comprehensive Community Initiative." *Urban Affairs Review* 36 (3): 291-323.

Cole, L. W., and S. R. Foster. 2001. *From the Ground Up: Environmental Racism and the Rise of the Environmental Justice Movement.* New York: New York University Press.

Dukes, E. F. 1996. *Resolving public conflict: transforming community and governance.* Manchester, U.K.: Manchester University Press.

Dukes, E. F. 2000. *Reaching for Higher Ground in Conflict Resolution: Tools for Powerful Groups and Communities.* San Francisco: Jossey-Bass.

Dukes, E. F. 2004. "What We Know about Environmental Conflict Resolution: An Analysis Based on Research." *Conflict Resolution Quarterly* 22 (1-2): 191-220.

Eisen, J. 1999. "Brownfields policies for sustainable cities." *Duke Environmental Law and Policy Forum* 9:187-230.

Elliott, M., and G. Bourne. 2005. *Evaluating the U.S. Environmental Protection Agency's Brownfields Facilitation Pilot Projects.* Atlanta: Southeast Negotiation Network.

English, M. R., A. K. Gibson, D. L. Feldman, and B. E. Tonn. 1993. *Stakeholder Involvement: Open Processes for Reaching Decisions about the Future Uses of Contaminated Sites.* Knoxville: Waste Management Research and Education Institute, University of Tennessee.

Fernandez-Gimenez, M. E. Forthcoming. "How CBCs Learn: Ecological Monitoring and Adaptive Management." In *Community-Based Collaboration for Natural Resource Management: Putting Knowledge to Work,* E. F. Dukes, K. Firehock, and J. Birkhoff, eds.

Fisher, R., and W. Ury. 1981. *Getting to Yes: Negotiating Agreement Without Giving In.* Boston: Houghton Mifflin.

Forester, J. 1999. *The Deliberative Practitioner: Encouraging Participatory Planning Processes.* Cambridge: MIT Press.

Foster-Fishman, P. G., S. L. Berkowitz, D. W. Lounsbury, S. Jacobson, and N. A. Allen. 2001. Building collaborative capacity in community coalitions: a review and integrative framework. *American Journal of Community Psychology* 29(2):241-

Greenberg, M., and M. J. Lewis. 2000. "Brownfields Redevelopment, Preferences and Public Involvement: A Case Study of an Ethnically Mixed Neighborhood." *Urban Studies* 37 (13): 2501-2514.

Henton, D., and J. Melville. 2005. *Collaborative Governance: A Guide for Grantmakers.* San Francisco: William and Flora Hewlett Foundation.

Holling, C. S., ed. 1978. *Adaptive Environmental Assessment and Management.* New York: John Wiley.

Innes, J. 1999. "Evaluating Consensus Building." In *The Consensus Building Handbook: A Comprehensive Guide to Reaching Agreement,* L. Susskind, S. McKearnan, and J. Thomas-Larmer, eds., 631-675. Thousand Oaks, Calif.: Sage.

Innes, J. 2002. *Evaluation Design for the Capacity Building Program of the California Center for Public Dispute Resolution.* Sacramento. California Center for Public Dispute Resolution.

Kettering Foundation. (1991). *Citizens and Politics.* Dayton, Ohio: Kettering Foundation.

Leach, W. D., and N. W. Pelkey. 2001. "Making Watershed Partnerships Work: A Review of the Empirical Literature." *Journal of Water Resources Planning and Management* 127 (6): 378-385.

Leach, W. D., and P. A. Sabatier. 2005. "To Trust an Adversary: Integrating Rational and Psychological Models of Collaborative Policymaking." *American Political Science Review* 99 (4): 491-503.

Lederach, J. P. 2005. *The Moral Imagination: The Art and Soul of Building Peace.* Oxford: Oxford University Press.

Lee, K. N. 1999. "Appraising Adaptive Management." *Conservation Ecology* 3 (2): 3.

Lukensmayer, C., and S. Brigham. 2005. "Taking Democracy to Scale: Large Scale Interventions for Citizens." *Journal of Applied Behavioral Science* 41 (1): 47-60.

Matsuda, M. 1995. "Looking to the Bottom: Critical Legal Studies and Reparations." In *Critical Race Theory: The Key Writings That Formed the Movement,* K. Crenshaw, N. Gotanda, G. Peller, and K. Thomas, eds., 63-79. New York: The New Press.

Pepper, E. N. 1997. *Lessons from the Field: Unlocking Economic Potential with an Environmental Key.* Washington, DC: Northeast Midwest Institute.

Scholz, J. T., and B. Stiftel, eds. 2005. *Adaptive Governance and Water Conflict: New Institutions for Collaborative Planning.* Washington, DC: Resources for the Future.

Susskind, L., S. McKearnan, and J. Thomas-Larmer, eds. 1999. *The Consensus Building Handbook: A Comprehensive Guide to Reaching Agreement.* Thousand Oaks, Calif.: Sage.

Teske, N. 2000. "A Tale of Two TAGs: Dialogue and Democracy in the Superfund Program." *American Behavioral Scientist* 44 (4): 644-678.

U. S. Environmental Protection Agency. 2001. *Superfund Community Involvement Handbook.* Washington, DC: Office of Emergency and Remedial Response.

U. S. Environmental Protection Agency. 2004. *Final Report.* Washington, DC: Superfund Subcommittee of the National Advisory Council for Environmental Policy and Technology.

U.S. Environmental Protection Agency. 2005. *FY 2004 Superfund Annual Report* (No. EPA-540-R-05-001). Washington, DC: U.S. Environmental Protection Agency, Office of Solid Waste and Emergency Response.

Walker, G. B., and S. Senecah. In preparation. "Collaborative Governance, Institutions, and Actors." In *Community-based Collaboration for Natural Resource Management: Putting Knowledge to Work,* E. F. Dukes, K. Firehock and J. Birkhoff, eds..

Wernstedt, K., and R. Hersh. 1998. "Through a Lens Darkly: Superfund Spectacles on Public Participation at Superfund Sites." *Risk, Health, Safety and Environment* 9 (2): 153-173.

Wernstedt, K., and K. N. Probst. 1997. *Land Use and Remedy Selection: Experience from the Field – the Industri-Plex Site.* (Discussion Paper 97-27). Washington, DC: Resources for the Future.

Wilkinson, J., A. B. Dotson, E. F. Dukes, and M. Hancox. 2003. *Central Chemical Superfund Redevelopment Initiative Pilot Project.* The City of Hagerstown, Hagerstown Land Use Committee, E2 Inc., University of Virginia's Institute for Environmental Negotiation (IEN).

Williams, B. L., K. S. Hoi, S. Brown, , R. Bruhn, R. de Blaquire, , and S. E. Rzasa. 2001. "Hierarchical Linear Models of Factors Associated with Public Participation Among Residents Living Near the U.S. Army's Chemical Weapons Stockpile Sites." *Journal of Environmental Planning and Management,* 44 (1): 41-65.

Yaffee, S. L. 2003. "Measuring Progress: A Guide to the Development, Implementation, and Interpretation of an Evaluation Plan." Ann Arbor: Ecosystem Management Initiative, School of Natural Resources, University of Michigan.

APPENDIX

PROCESS TYPE	GOALS	SPECIFIC EXAMPLES
Information Distribution	• Provide information	• Web Sites • Press Releases • Newsletters
Meetings and Hearings	• Provide information • **Hear concerns** • **Build shared understanding**	• Public Hearing • Public Meeting
Workshops	• Provide information • Hear concerns • Build shared understanding • **Develop ideas/recommendations**	• Visioning Workshops • Design Charrettes
Advisory Groups	• Provide information • Hear concerns • Build shared understanding • Develop ideas/recommendations • **Build relationships**	• Community Advisory Groups (CAGs)
Consensus-building	• Provide information • Hear concerns • Build shared understanding • Develop ideas/recommendations • Build relationships • **Make decisions**	• Mediation • Collaborative Problem Solving

Less Interactive, Authority-Driven ⟷ *More Interactive, Shared Authority*

Figure 7.2. A spectrum of community involvement processes.

8

TOXIC SITES AS PLACES OF CULTURE AND MEMORY:
Adaptive Management for Citizenship

Daniel Bluestone
University of Virginia School of Architecture

INTRODUCTION

The urgent effort to clean and reclaim blasted landscapes – EPA Superfund sites and other polluted brownfields – often involves an unfortunate exercise of cultural and historical amnesia. The sites are cleaned of their toxic substances but they are also scrubbed of their history. This need not be the case. If former buildings and landscapes on Superfund sites were adapted to new uses and interpreted for the public, rather than destroyed during redevelopment, we would retain an important material framework for better understanding both the sites themselves and their surrounding communities. Moreover, with tangible traces of former uses left in place, we would have an important venue for learning about the human use, abuse, and stewardship of the built and natural landscape. On Superfund and brownfield sites where traces of industrial use and pollution are removed entirely, the landscape makes less sense to residents and visitors alike. People whose lives and livelihoods were bound up with these places lose important landmarks from their locality. We have pursued cleanup and redevelopment policies that fail to recognize the power and possibilities of the historical memories that hover over these sites.

Whether cleanup involves cap and cover methods where toxic materials are contained or neutralized on a site, or hog and haul methods where pollutants are dug up and moved elsewhere (perhaps to become someone else's problem), the outcome is generally the same – sites are delivered for redevelopment devoid of any physical trace of their history, of their pollution, or even of their cleanup. These approaches to remediation ignore the potential of site interpretation to foster a more vital politics of place (Kemmis 1990; Basso 1996; Kaufman 2001). On blasted landscapes, a politics grounded in site history would facilitate fuller and more informed community participation in the remediation process as well as in the decisions about the future uses of reclaimed sites.

This chapter surveys developments in industrial history and historic preservation that can be usefully drawn upon in ongoing remediation work. It will also scrutinize recent work on Superfund sites that reveals both the problems and possibilities of cultivating history and memory as part of the cleanup process.

On a Superfund site, fostering public amnesia regarding site history may at first glance seem a prudent means to promote economic reuse or redevelopment. Some people with an understanding of these polluted sites might feel that over time, the less said, the less shown, the better. Nevertheless, such blindness to the past is problematic and ultimately undercuts the very work we are trying to accomplish in the remediation of polluted sites. People involved in the cleanup could make their efforts more comprehensible and decidedly less scary to the public if they could reveal how the flows of materials and pollutants on toxic sites had actually taken place. To see toxic sites as part of a broader industrial process with material inputs, products, and by-products that all worked their way through the buildings and the site would promote a kind of critical understanding of basic site processes that could in turn lay the groundwork for understanding something of the processes of site pollution, site remediation, and site reuse. This general approach to industrial site history, although rarely applied to Superfund sites, has been honed by historians and industrial archeologists over the past three decades.

One of the great advantages of interpreting site history, pollution, and cleanup on toxic sites comes from foregrounding human agency and helping citizens reflect more critically both upon the past and the present. Site histories, circulated in the form of brochures with illustrations and diagrams, on web sites, and in public lectures at community meetings, can demystify the historical decisions whereby people interacted with each other, with the economy, and with the natural environment as they gave form to the human landscape, including its pollution and subsequent cleanup. Critically understanding history and plans for remediation on a polluted site helps us situate our own actions as linked in a profound way with the actions of citizens who came before us and who will come after us. In 1889, Harvard professor Charles Eliot Norton's essay "The Lack of Old Homes in America" recognized the ways in which architecture and landscape provide a palpable means for taking measure of human agency and values across the generations. Norton bemoaned the lack of old and hereditary residences in the United States. He worried about the social effects of transience on both public and private life. He wrote, "No human life is complete in itself; it is but a link . . . in a chain reaching back indefinitely into the past, reaching forward indefinitely into the future. Whatever weakens the sense of its linked relation is an evil" (1889: 638). It would be useful today to keep this insight in mind as we hone our approaches to managing Superfund cleanup.

The erasure of history on Superfund sites parallels the evil that Norton identified in the transience of people and families in the domestic landscape. Using a modern metaphor to push forward his analysis, Norton wrote, "To strengthen its connection in both directions, to quicken the electric current of conscious existence conveyed from the past through the present to the future, is to increase the vital power of the individual, his sense of dignity, and of responsibility. To the future every man owes the immeasurable debt for which he stands indebted to the past" (638). In grappling with the profound public health menaces on polluted sites, we, too, would benefit from a sense of the ways in which our actions are prescribed by people who came before us while defining the social, cultural, and economic possibilities of people in the future. The material remains of industrial processes and pollution, captured in existing buildings and landscapes, provide a powerful venue for taking the measure of such histories. This historical understanding in turn creates a firmer basis for our own action as citizens and residents caught up in site remediation. With the site as the focal point, history shades into action and helps constitute an informed politics of place.

Local and everyday landscapes (like those we deal with on Superfund sites), for their sheer familiarity and accessibility, are the places most capable of stirring an understanding of human agency and throwing into higher profile our own responsibilities as community members and citizens to the past and the future (Kemmis 1990). These blasted landscapes can again play a productive role in community by helping us constitute a politics attached to place, so that we can take on the challenge of toxic sites and our own ecological relationship to the natural environment. It is precisely these local histories and landscapes that often resonate most deeply with people who are increasingly buffeted by the growing placelessness and homogenization of the modern world, a world that obfuscates human agency and trivializes citizen action.

THE FRESNO SANITARY LANDFILL

In the summer of 2001, controversy erupted in Washington, D.C. and Fresno, California when Secretary of the Interior Gale Norton designated the Fresno Sanitary Landfill as a National Historic Landmark. A trash mound 4,200 feet long, 1,250 feet wide, and 45 to 60 feet high that was created between 1937 and 1987 joined Faneuil Hall, Boston's Old State House, and George Washington's Mount Vernon on a list of fewer than 2,500 American places with the coveted designation of a National Historic Landmark (see Figure 8.1) (Melosi 2002). These are the places that rise to special national status from the nearly 80,000 listings on the National Register of Historic Places. I want to briefly review the Fresno history because it underscores the challenge of recognizing and disseminating information about the everyday landscape.

Figure 8.1. Fresno Sanitary Landfill aerial view, c. 2003.
Courtesy City of Fresno.

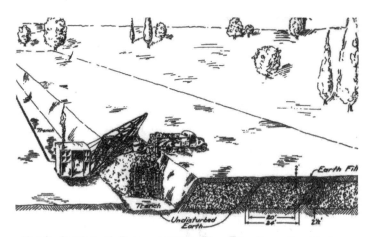

Sketch drawing of Fresno sanitary fill. This disposal method is being used by a number of other cities and at Army camps.

Figure 8.2. Sketch of Fresno Sanitary Landfill Operation, c. 1945.
Courtesy National Park Service.

It is notable that the designation of the Fresno dump would have been unthinkable in the 1960's when the National Register program started. Like the study of the history of architecture and the built landscape, to which it is closely linked, historic preservation has expanded dramatically—increasingly surveying, documenting, designating, and interpreting the history and forms of everyday life. In this way, historic preservation has come to embody the popular and populist insights of social history and complement the earlier patriotic, nationalist, and aesthetic basis for historic preservation (Wallace 1996a, 1996b; Bluestone 1999). The Fresno Sanitary Landfill did not loom large in the political history of the United States, nor did it have an especially significant aesthetic aspect. Nevertheless, when it was established in 1937 it represented an innovative approach to the handling of municipal solid waste. Long trenches between 20 to 24 feet wide and 10 to 35 feet deep were dug at the site. The trenches were then filled with garbage, leveled, compacted, and covered with dirt (see Figures 8.2 and 8.3). The guiding idea promoted by Fresno municipal engineer Jean Vincenz was that the layering and compaction of dirt would help control rodent problems and reduce the volume of debris. This method eventually won broad favor over the then current practice of dumping garbage on open land or in water or incinerating it, fouling the air and still leaving piles of ash for subsequent disposal (Melosi and National Park Service 2000). It soon became a model for municipalities around the country.

Figure 8.3. Garbage disposal at Fresno Sanitary Landfill, 1935.
Courtesy National Park Service.

Vincenz's method and the leading role that the Fresno site played in popularizing the trench and compaction methods of disposal provided the rationale for listing the site as a National Historic Landmark.

The designation of the Fresno site grew out of a conscious effort to "broaden the scope" of the National Historic Landmark program (Melosi 2002: 21-22). The National Park Service worked with Professor Martin V. Melosi, director of the Institute of Public History at the University of Houston, and a historian of municipal sanitation and infrastructure, to carry out the research that led to the nomination of the Fresno Sanitary Landfill as a Historic Landmark (Melosi 2002). Melosi succeeded in moving the National Historic Landmark program into new and uncharted landmark territory—into an area that he had helped develop as a field of historical inquiry (Melosi 2000). The effort seems entirely salutary, for at its base the handling of garbage is a fundamental and telling aspect in the history of human society. Coming to terms historically with sites like the Fresno Sanitary Landfill has tremendous possibilities not only for historical insight but also for helping us focus on our ongoing relationship to natural resources, waste, and the environment. It is difficult to survey the history of a monumental site like the Fresno Landfill without moving our perspective forward to contemplate our own piles of garbage and our own systems of disposal.

Still, for some the designation of the Fresno site seemed evidence of preservation run amok—it simply didn't fit commonly held notions of preservation and its link to the canon of nationalism or aesthetic accomplishment. Paul Rogers wrote in the *San Jose Mercury News*, "Other presidents have honored Pearl Harbor, Alcatraz and Martin Luther King's birthplace as national historic landmarks. Now the Bush administration has added its own hallowed place: a garbage dump in Fresno" (Rogers 2001, quoted in Melosi 2002: 19). Despite the merits of the designation, the criticism of Interior Secretary Gale Norton prompted her to quickly backtrack, seeking to dump the dump by "temporarily" rescinding the historical designation. Residents of Fresno were not at all sure they wanted to be known nationally for their dump. Environmentalists seized upon the designation as an ironic tribute to an administration with decidedly questionable commitments to the environment, which had decided to designate a dump as a historic site rather than protecting remaining stands of giant sequoia trees or coastlines eyed for oil drilling.

What was even more controversial about the designation was the fact that the Fresno site had for years accepted battery acid from a local manufacturer, medical wastes from hospitals, and other toxic substances. The designers of the Fresno Sanitary Landfill had certainly not anticipated that decomposition of garbage would eventually create dangerous levels of methane gas that would migrate off the site or that volatile organic chemicals would turn up in the region's groundwater. In 1987, the EPA ordered the closure of the Fresno

Landfill. In 1989, the EPA placed the property on the National Priorities List of Superfund sites, among the most dangerously polluted sites in the country. The site was then cleaned up by the EPA at a cost of 9.5 million dollars. Equipment to collect and flare methane gas was installed and a geomembrane liner and four to five feet of additional soil was placed over the entire trash mound to prevent water run-off from polluting the area. Still, much political theater was made possible by the fact that at a time when the Bush administration was withdrawing from international environmental treaties it was designating as a National Historic Landmark a landfill that had been a Superfund cleanup site (Melosi 2002; Grossi 2005).

Secretary Norton announced the rescinding of the Fresno designation on August 28, 2001. September 11th turned attention to other political matters and in its aftermath Norton never executed the written memo to rescind her designation. Today the Fresno Sanitary Landfill still occupies a spot on the National Historic Landmarks list.

THE NEED TO ENGAGE INDUSTRIAL SITES HISTORICALLY

What I would like to consider is the extent to which the landmark designation has been disseminated to a broad public. Public awareness of the site history was clearly spurred when the designation debate played out in newspapers and on television in August 2001, but interest seems to have quickly passed after September 11th. What I find most striking is the fact that the public authorities who control the site have made no effort to share its history with the thousands of people who now visit the site every week. In 2002 Fresno built a 350-acre regional sports park in the area, including the landfill itself. The new park has five playgrounds, six softball fields, and seven soccer fields (see Figure 8.4). The web site for the city parks calls the area an "environmentally conscious facility, which includes integration of a former landfill site." There are also hiking trails and a "hilltop overlook" (City of Fresno 2006).

But for people using those trails or surveying the scene from the overlook, there is nothing on the site – no marker, no brochure, no historical photographs, and no personal interpretations – that present the significant history of the Fresno Sanitary Landfill. There is nothing that points out that the "hilltop overlook" occupies an eminence created with 7.9 million cubic yards of people's garbage deposited over a period of half a century (Melosi and National Park Service 2000). There is no design of historical engagement with the various human processes that either polluted or cleaned up the site. No one is drawn into a consideration of where the piles of bottles and soda cans that collect in the Sports Park trash receptacles are going to end up. Expanding the canon of National Landmarks seems all very well and good;

however, we should ask why there is no room in a 25 million dollar budget for a sports park to disseminate the site's rich history and to encourage reflection about the process and politics of garbage and human relationships to natural resources and the environment (see Figure 8.5)?

Figure 8.4. Fresno Sports Park with capped landfill at right, c. 2003.
Courtesy City of Fresno.

Figure 8.5. An Unexplored Audience of Athletes and Fans at Fresno
Sports Park, 2003.
Courtesy City of Fresno.

As a Superfund site the Fresno Sanitary Landfill today has engaged in precisely the sort of cultural and historical amnesia that has become such a common element of other toxic cleanup sites. The public should not have to read scholarly journals or back issues of local newspapers to engage the history of a site that is so ripe for interpretation. The site should have been designed, marked, and interpreted in a way that made the history of the "overlook" mound more palpable. Modern sports parks today often have parcours fitness trails – mile-long walking or running courses that connect 15 to 20 exercise stations with apparatus and instruction signs for completing various exercises. These systems provide a framework for individual exercise and health (Knudson 2005). At Fresno a similar system with a path and individual interpretative stations that explored the history of the landfill and the site remediation could provide a framework for the social exercise of citizenship, for civic and public health education. Signs atop the "overlook" could declare that you are now standing on a pile of 7.9 million cubic yards of trash. Information could be provided about the amount of waste generated by the average household today and where that garbage ends up. Another sign or station might explore the public health danger of the toxic pollution caused by the landfill before it was contained and cleaned up. Another sign could describe the remediation technology that permitted the site to be used as a sports park; it could explain the methane being flared off at the site today. Working more directly with traces of the site's history could begin to constitute a politics of place in Fresno whereby people would understand the history of waste while being encouraged to think about their own levy on the land and its resources. This is a project on which historians, designers, and ecologists as well as planners could productively collaborate.

Historians and designers certainly have a stake in environmental cleanup, in confronting the many threats posed by polluted toxic environments. However, beyond this broad citizenship interest, it seems that we can quite productively bring to bear the methods and insights of our disciplines to better confront the challenges of toxic waste sites. A key ideal of many Superfund and brownfield reclamation projects is not only to mitigate environmental problems but to recycle or reuse abandoned sites that are often already tied into existing transportation and utility infrastructures. I would argue that these sites are also tied into an infrastructure of community history and memory (Mason 2004); exploring such a perspective offers tremendous potential to engage the work of citizens, communities, and agencies working for site reclamation and renewal.

As in the case of the Fresno designation, the idea of establishing some alignment between historic preservation and environmentally toxic sites might appear fraught with difficulty. The EPA's health and environmental goal is to remove or contain toxic substances in a way that seems very much at odds with preservationists' efforts to develop site-specific narratives of place and to engage people directly in the material dimensions of history on a site. Preservationists aim to recognize, frame, and chronicle the material traces of history in the landscape. Preservation work often focuses on restoration and renovation. In a project of adaptive reuse, preservationists aim to find new uses for buildings and sites that take the place of the obsolete uses for which they were originally designed. People charged with the cleanup of toxic sites aim to eradicate or at least to neutralize the accumulations of the past. The initial effort is, of course, to eradicate pollution, but the buildings and landscapes tied to the pollution are often casualties of the process. This opposition between preservation and remediation seems quite unfortunate.

As we think about the political and citizenship dimensions of designing with and interpreting history on Superfund sites, it is important to keep in mind that there is a long tradition of political calculus accompanying historic preservation projects. In 1850, in one of the earliest acts of public preservation in the United States, the New York State Legislature purchased Washington's revolutionary war headquarters at Newburgh, New York. The committee of the legislature argued that the physical properties of buildings and landscapes made their associated histories especially tangible, capable of "transmit[ting] to our children a knowledge of the virtues of the fathers of the republic." The committee declared: "If our love of country is excited when we read the biography of our revolutionary heroes, or the history of revolutionary events, how much more will the flame of patriotism burn in our bosoms when we tread the ground where was shed the blood of our fathers, or when we move among the scenes where were conceived and consummated their noble achievements" (Caldwell 1887: 21). In the 1850's, preservation of this headquarters seemed especially important. Here, in 1783, Washington stemmed a rising revolt among his own troops who were organizing to march on the Congress because they had not been paid for their military service. The preservationists had a very clear political purpose; they wanted to diffuse portraits of nationalism and the memory of Washington's leadership to defuse the growing tensions over slavery. They wrote, "It will be good for our citizens in these days of political collisions, in these days of political demagoguism: it will be good for them in these days when we hear the sound of disunion reiterated from every part of the Country; in all future time occasionally to chasten their minds by reviewing the history of our revolutionary struggle" (1887: 21).

The preservation of the headquarters was more successful than the political agenda that prompted it. Nevertheless, this early preservation campaign revealed that ideals of politics and citizenship often complemented the curatorial zeal that seemed to stand at the center of the preservation movement. It is this effort to make use of sites in framing a politics of place that can usefully be adapted to the cleanup and redevelopment of toxic sites.

CULTIVATING A POLITICS OF PLACE TO ENCOURAGE SITE STEWARDSHIP

In aiming to cultivate a politics of place on Superfund sites it is important to realize that the industrial landscape has historically lacked the recognition afforded to sites associated more directly with politics and nationalism or those celebrated for their aesthetic significance. This began to change amidst the social movements of the 1960's, which inspired different approaches to history and a new interest in more varied architectural and urban forms. Social history, labor history, and the history of vernacular architecture and everyday landscapes each flowed from an effort to move beyond the established canon of political history and the history of social and economic elites, and in architectural history, beyond high style architecture. Historians grew increasingly adept at chronicling the history of everyday life, or working people, or people outside of the circles of privilege associated with political, social, and economic elites. The federal government had started recording historic buildings in 1933 under the auspices of the National Park Service's Historic American Building Survey. In 1969 the Park Service established the Historic American Engineering Record to carry out the systematic documentation of engineering and industrial sites—including textile mills, bridges, dams, and other utilitarian structures. Industrial sites offered a new and very important perspective on American social and economic history. The Society for Industrial Archeology was founded in 1971 to promote the study of industrial buildings and sites. Preservationists also saw in vacant or under-utilized industrial sites a substantial stock of buildings that could be adapted to new residential or commercial uses. Thousands of new units of housing, many for senior citizens, filled the long vacant spaces of 19th and early 20th century New England textile mills and warehouses. The conversion of the Ghirardelli Chocolate Company's San Francisco factory in the early 1960's into shops and restaurants underscored a growing public interest in industrial history and buildings. The loft buildings of New York's Cast Iron District in SOHO assumed an entirely new meaning as trendy residential spaces. The unraveling of consensus over the actions of political and economic elites in the 1960's was partly reflected in a parallel shift in the priorities of historians, preservationists, and the general public. The history of industrial sites and working class life gained considerable prominence during the 1960's and 1970's (Murtagh 2006).

Scholarly and public interest in industrial heritage has spurred the development of industrial history and increased the sophistication with which we are able to scrutinize the dynamic relationship between people, technology, buildings, sites, and the broader economy. Observers, drawing on the work of historians of technology and labor and industry, are able to engage industrial sites historically with a fairly high degree of rigor. Using the historical methods of industrial, labor, and environmental history it is possible at these sites to come to terms with the important intersection between resources, labor, technology, and the environment to understand precisely how industries operated in buildings and on the land as well as within broader systems of transport, finance, and consumption (Cronon 1991). Today, these same methods can be used to better come to terms with the architectural, cultural, economic, and environmental history of polluted industrial sites. The history of many American communities is inextricably linked to the industries that help explain a central part of their very existence. The industrial landscape has great potential in helping people take the measure of local and regional history. This is the case whether the sites are polluted or not.

In 2003, my colleague Julie Bargmann and I and our students in the School of Architecture at the University of Virginia explored ways of drawing upon the methods and framework of industrial history in narrating the history of, and designing reuse proposals for, a Superfund site in Hagerstown, Maryland. We undertook a yearlong project that focused on Hagerstown's Central Chemical Superfund site (see Figure 8.6). The work of our course looked at systems such as industrial and environmental flows both on the site and in the broader context. It quickly became apparent that Superfund sites could usefully be approached both historically and in terms of their future redevelopment in a broader context. Just as the original industrial development relied on adjacent industries, workers drawn from the town, and the structure of transportation and investment, revealing these connections lays the basis for a more transparent approach to the future possibilities and limitations of redevelopment. An approach that builds on and cultivates a cultural, historical, and geographical perspective seems best calibrated to raise the most important questions about future redevelopment. Many of the places with Superfund sites are dealing with pervasive de-industrialization. An analysis of the history of Superfund sites can give citizens an important framework for coming to grips with the broader economic trends that continue to shape their communities.

An additional advantage of an emphasis on keeping, identifying, interpreting, and reusing major elements of a Superfund site's former industrial occupation is that it promotes a more gradual, incremental approach to the site that will tend to cultivate mixed-use scenarios for reuse. The redesign is more likely to breed some complexity as the future program is adjusted and integrated with existing buildings and spaces. Redesign efforts that focus on the reuse of existing industrial structures on the site tend to encourage a more interesting set of designs that envision complex, multi-use, chronologically disparate approaches to redevelopment. This sort of redesign is quite likely to be more useful to a variety of stakeholders in the surrounding community. Our students' designs for the Central Chemical site took

Figure 8.6. Central Chemical buildings and site, 2003.

seriously the existing buildings and the varying levels of pollution across the site. This work contrasted with the penchant for single-use redevelopment schemes that place a soccer field, a warehouse district, or a new shopping center on reclaimed sites. It was precisely the varied and mixed-use palette of our proposals that prompted the greatest enthusiasm from Hagerstown's residents, who had been struggling through the process of envisioning the future of the Central Chemical Company site. Here adaptive reuse went beyond preservation's interest in finding new uses for existing buildings and sites; it actually set out to foster environmental stewardship and to enrich the lives of citizens engaged in both site remediation and subsequent reuse.

As our project developed, we had the opportunity to present our histories and our designs to many people in the community. It was readily apparent that important forms of social capital were created through a process that helped people see their Superfund sites and their communities in palpable historical terms. As a Superfund site and its broader community come into focus historically, people's understandings and attachments to them tend to strengthen. We provided guidebooks, exhibition posters, and a web site that explored aspects of Hagerstown's industrial and architectural history. At the Central Chemical site, knowing that in the 1950's the firm began manufacturing the pesticide DDT and that it dumped batches into a pond on the site that drained into Antietam Creek and then the Potomac River helps delineate the bounds of the problem. Historical understanding and connection pave the way toward a more informed public involvement and stewardship of the site and the future of the community. By grounding the cleanup in public historical understanding, remediation is not simply the province of experts seeking detachment from an angry or scared citizenry. Rather, it is a process grounded in some clarity about both history and future possibilities.

History is one important element in constituting a politics of place. It is the historical perspective that foregrounds the dynamic of human agency and lets people move more or less seamlessly from an understanding of historical agency on a particular site to a deeper conception of their own actions as citizens. In the case of Superfund sites, design that reveals the process of environmental remediation constitutes another important venue for strengthening understanding and the sort of ongoing stewardship of these sites that they demand. What Julie Bargmann and I have done is to encourage design and interpretative strategies that will bring people into a more profound understanding of polluted sites. We have proposed first using industrial site histories to help people better understand the issues raised by polluted sites. These can take the form of discussions in public meetings, web sites, information brochures, exhibitions, guided tours of sites, and site markers.

We have also suggested that designers incorporate the traces of history into the designs for reuse of these sites in order to preserve a palpable venue for ongoing site interpretation. Students in our Hagerstown studio project developed creative ways of incorporating the physical traces of industrial production and environmental remediation into their design proposals for the reuse of the Central Chemical site. Gretchen Kelly Giumarro designed a project that she termed "Industrial Nesting." It was founded on the idea that DDT not only fouled the Central Chemical site but also inhibited the absorption of calcium by various species of birds (leading them to lay weak-shelled eggs), drastically reduced the reproduction rates of bald eagles, osprey, and peregrine falcons. Giumarro proposed a diverse bird habitat for the site. Rather than removing the buildings from the site, she proposed that the buildings, with the link to the history of industry and pollution on the site, be converted into bird perches and blinds for birdwatchers (See Figure 8.7).

Figure 8.7. Gretchen Kelly Giumarro, "Industrial Nesting,"
Hagerstown BOOM Studio, 2003.

Other buildings were also converted into educational and recreational centers. This "redemptive" design proposal maintained the site history, giving it a meaning and significance far greater than if the site were simply developed as a bird habitat with no connection to the buildings, landscapes, or history of Central Chemical. Brian Gerich drew inspiration from another aspect of Central Chemical's history of pollution—the fact that its water run-off polluted local streams. He developed a phased plan whereby the site would take on a larger and larger role in gathering neighborhood storm water and filtering it as the remediation process advanced (See Figure 8.8).

Figure 8.8. Brian Gerich, "Watershed Park,"
Hagerstown BOOM Studio, 2003.

Buildings on the site would provide incubator spaces for companies doing research in ecological technologies. Kent Dougherty designed a sports park that rather than hiding the history of local contamination actually used the bio-piles and quarries developed during site remediation and gave them a new use as a foundation for a skateboard, in-line skating, and biking park. Dougherty's design departed from the usual approach of flattening the site for redevelopment and removing the traces of remediation (Figures 8.9 and 8.10).

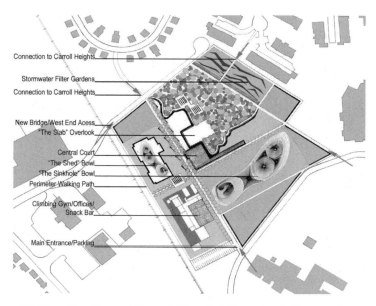

Figure 8.9. Kent Dougherty, "Central Chemical Sports Park,"
Hagerstown BOOM Studio, 2003.

Instead, the landforms created during remediation were simply given a new use, interpreted through signs and exhibits. Cara Ruppert designed an "EcoLab Park" where as remediation progressed, a center for studying urban ecology and pollution, a laboratory, a library, and an educational building would be established in the old Central Chemical structures (Figure 8.11).

Figure 8.10. Kent Dougherty, "Central Chemical Sports Park,"
Hagerstown BOOM Studio, 2003.

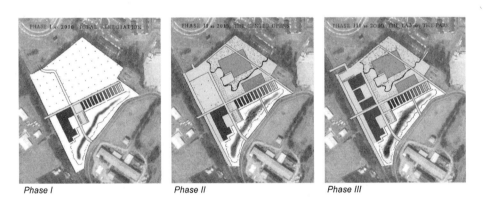

Phase I   Phase II   Phase III

Figure 8.11. Cara Ruppert, "Central Chemical EcoLab Park,"
Hagerstown BOOM Studio, 2003.

Historical exhibits in those buildings would convey the full site history in palpable terms—represented by the reused buildings. The site would also take a hand in storm water filtering, reversing the history of pollution on the site. Ben Spencer envisioned mixed-uses for Central Chemical buildings and grounds, including agriculture, wildlife habitat, a renewable energy plant, a recycling center, and a public market (Figure 8.12).

Figure 8.12. Ben Spencer, "Central Chemical Farm,"
Hagerstown BOOM Studio, 2003.

Sarah Trautvetter designed a dynamic landscape that envisioned the gradual return of people to the Central Chemical site—this time as a cultural center and outdoor theater (Figure 8.13). As the site was cleaned the fences would be moved further into the park and the spectacle of performance in the park would go from "standing room only," observing the remediation process from afar, to full audience participation in cultural events. The strength of these proposals resided in part in their cultivation of a history and a politics of place rooted in the specific buildings and landscapes and associations of the Central Chemical plant. Such engagements were largely absent from the earlier re-development proposals, which like the treatment of the Fresno site completely ignored the power and possibilities embedded in the historic landscape, and open to interpretation by historians, ecologists, designers, and local residents.

Many of the Central Chemical projects included remediation as part of the site history that was framed and made visible. In going public with site histories we can foster a firmer foundation for citizens struggling to envision the future shape and operations of their community—visions that are best informed by a more profound understanding of the intricate connections between human knowledge, values, and power and the actual form of settled human landscapes. In this sense, while working toward novel ends, we work within the older political framework of preservation evident in the campaign to preserve Washington's Headquarters. The mid-19th century preservationists promoted an understanding of place that was historical; however, in their view its importance rested in no small part on motivating future political action by citizens who derived some of their politics and understanding of the world from tangible encounters with what Dolores Hayden (1995) has termed "the power of place." For historians, preservationists, designers, and the public, such projects can spur an important dialogue that links insights about the past with visions for the future.

Figure 8.13. Sarah Trautvetter, "Landscape Theaters,"
Hagerstown BOOM Studio, 2003.

CONCLUSION

Our approach to industrial sites and the effort to use site interpretation to make citizens more aware and engaged in pressing issues related to the future form and nature of community life is part of the broader effort to come to terms historically with everyday landscapes. This process would not have been possible without the sustained commitment on the part of a generation of historians to grapple with the form and meaning of everyday landscapes. In this sense I very much appreciate Joseph Amato's advocacy of local histories in his recent book *Rethinking Home: A Case for Writing Local History* (2002). Amato argues that

> local historians provide a passionate attachment to concrete places in an age when home and place, locale and landscape, are in a state of great mutation. This tension provides the basis for an ever deepening conversation. . . . [Local history] awakens a passion for understanding the compass of local action. In this way, local history serves the intelligence that frees the energy of local people to work in the dimensions of the possible. Committed to understanding the present and the changes that characterize it, local history proves a golden asset for all vital people of a place.

It galvanizes citizenship.

Serious efforts to understand the form and history of the everyday landscape on Superfund sites give us an ideal venue for working with a broad popular public. History is no longer something set apart, detached, captured in studies of the houses of a wealthy elite who could command the services of famous architects. We can now begin to take seriously the everyday places where most people live, work, and visit. The challenge is getting the insights out of the academy and disseminated in more public and popular venues—in designs for the renewal of Superfund sites, in historic districts, in guidebooks of neighborhood history and architecture, on web sites that feature the Superfund remediation process for specific sites, in museum exhibitions, in secondary school curricula that focus on local history and architecture, and in popular journalism and lectures. Lots of good work has been done in this regard. Approaching toxic sites as places of culture and memory is a politically important and intellectually vital project that should be expanded.

REFERENCES

Amato, Joseph A. 2002. *Rethinking Home: A Case for Writing Local History.* Berkeley and Los Angeles: University of California Press.

Basso, Keith H. 1996. *Wisdom Sits in Places: Landscape and Language Among the Western Apache.* Albuquerque: University of New Mexico Press.

Bluestone, Daniel. 1999. "Academics in Tennis Shoes: Historic Preservation and the Academy." *Journal of the Society of Architectural Historians* 58 (September): 300-307.

Caldwell, Richard. 1887. *A True History of the Acquisition of Washington's Headquarters At Newburgh, By the State of New York.* Middletown, NY: Stivers, Slauson & Boyd.

City of Fresno. 2006. http://www.fresno.gov/parks-rec/parkdisplay.asp?RecNo=117

Cronon, William. 1991. *Nature's Metropolis: Chicago and the Great West*. New York: Norton.

Grossi, Mark. 2005. "Dump Cleanup Inspected, " *Fresno Bee,* 19 March.

Hayden, Dolores. 1995. *The Power of Place: Urban Landscapes as Public History*. Cambridge: MIT Press.

Kaufman, Ned. 2001. "Places of Historical, Cultural, and Social Value: Identification and Protection, Part I." *Environmental Law in New York* 12 (November): 211-233.

Kemmis, Daniel. 1990. *Community and the Politics of Place*. Norman: University of Oklahoma Press.

Knudson, Douglas. 2005. Park Trails for Fitness. Available at http://www.ces.purdue.edu/extmedia/FNR/FNR-106.html. Accessed 2 September 2005.

Mason, Randall. 2004. "Historic Preservation, Public Memory, and the Making of Modern New York City." In *Giving Preservation a History: Histories of Historic Preservation in the United States,* edited by Max Page and Randall Mason, 143-157. New York: Routledge.

Melosi, Martin V. 2000. *The Sanitary City: Urban Infrastructure in America from Colonial Times to the Present*. Baltimore: The Johns Hopkins University Press.

Melosi, Martin V. 2002. "The Fresno Sanitary Landfill in an American Cultural Context." *Public Historian* 24 (Summer): 17-35.

Melosi, Martin V., and the National Park Service. 2000 (August). "Fresno Sanitary Landfill," National Historic Landmark Nomination, National Park Service. Washington, D.C.: Department of Interior.

Murtagh, William J. 2006. "Rehabilitation and Adaptive Use." In *Keeping Time: The History and Theory of Preservation in America*, 99-106. New York: John Wiley & Sons.

Norton, Charles Eliot. 1889. "The Lack of Old Homes in America." *Scribner's Magazine* 5 (May): 638.

Rogers, Paul. 2001. "Leaky Trash Site Chosen as National Treasure." *San Jose Mercury News,* 28 August.

Wallace, Mike. 1996a. "Preserving the Past: A History of Historic Preservation in the United States." In *Mickey Mouse History and Other Essays on American Memory,* 178-210. Philadelphia: Temple University Press.

-----------------. 1996b. "Preservation Revisited." In *Mickey Mouse History and Other Essays on American Memory,* 224-237. Philadelphia: Temple University Press.

9

CHAT:
Approaches to Long-Term Planning for the Tar Creek Superfund Site, Ottawa County, Oklahoma

Niall G. Kirkwood
Harvard University Graduate School of Design

> *I am worried about the people in the Tar Creek area, but mostly for all the little children. Children are so vulnerable, that's why we have to teach them to wash their hands, not to play in Tar Creek, and what lead is. I am against moving the towns. There is so much history there and compassionate people that care for their town. My mother is the postmaster at the Picher Post Office and I know that there is no other place she would rather be than Picher. She loves the people in the town and doesn't want them to move, or her for that matter.*

Samantha Proctor, 11th Grade
Miami High School, Miami, Ottawa County, Oklahoma (CYC 1999)

INTRODUCTION

Tar Creek in Ottawa County, Oklahoma, a former lead mining area and part of the Tri-State Mining Area of Oklahoma, Kansas, and Missouri, represents the failure at multiple levels of administration and overview to address decades of environmental degradation, toxic land conditions, and environmental injustice. Communities and environmental officers, in collaboration with local agencies and academic research groups, have started to address the long-term remediation, planning and regeneration of vast mining soil heaps, polluted waterways and rivers, and land subsidence and mineshaft sinkholes that characterize the Superfund site's landscape and surrounding waste territories (see Figure 9.1).

Figure 9.1. Rebecca Jim of LEAD surveys a chat pile.

The story of the long-term reclamation of the forty square mile Tar Creek Superfund Site is presented here as a case study of future planning and design strategies that utilize the resources of local residents, stakeholders, and environmental organizations with the assistance of academic research institutions. I will argue that local-based transformations of the site, as opposed to "top-down" actions, have the potential to act as a framework to unite community desires, address environmental cleanup, and regenerate sequentially, over time, the entire landscape of a polluted area.

This chapter also argues for a holistic way of viewing Superfund sites that acknowledges and then acts on the need for an incremental approach to land reuse and change within existing local social, cultural, and ecological environments. It challenges the strategies that are currently brought to bear on these sites through the existing Superfund law and focuses on the complexity of these landscapes as places with attendant histories, ecologies, communities, and potential futures. It describes the nature, built fabric, and materiality of these places and the normative and often simplistic means that have been used to alleviate their current conditions of degradation and contamination. In the end, it is the people, and the physical landscapes that they shape, restore, or reclaim for their daily uses and future utility, who are the inhabitants of sites such as Tar Creek, and that we turn to for guidance.

As cultural critic John Berger has written (Berger & Mohr 1981), and as could be applied to many of the Superfund sites in our contemporary environment:

> Landscapes can be deceptive –
>
> *sometimes a landscape seems to be less a setting*
> *for the life of its inhabitants than a curtain behind which*
> *their struggles, achievements and accidents take place.*

As outsiders, we have to be careful when thinking about a landscape, whether polluted or not, that we know a place and its people. This is especially true for those who have the responsibility to address the serious environmental concerns at Superfund sites on behalf of federal, state and local agencies or other interested municipal institutions. The focus of this work is broadly concerned with the reclamation of derelict land, where the cleaning of land, water, and soils becomes a sustainable infrastructural investment for the communities and stakeholders rather than just a selected environmental technique to be applied to a polluted site. The former mining works at the Tar Creek Superfund Site in northeast Oklahoma – a Superfund site that, although it has been under scrutiny since the early 1980's, is still in its very early days of recovery and planning – serves as a useful model for this approach.

TWO SETS OF PHOTOGRAPHS

Nancy Goldenberg grew up and went to grade school within the Tar Creek Superfund Site in Ottawa County, Oklahoma. She has continuously returned from Chicago where she now works as a professional photographer to document the landscapes of Tar Creek, her home landscape. From the vantage point of a light aircraft rented by the hour, she has photographed the local communities, businesses, roadways, homes, and her former grade school within an expansive landscape of mining waste (or "chat"), threaded by orange-red contaminated creeks, rivers, and ponds, and littered with collapsed sinkholes as shown in Figure 9.2. These and similar images (also documented by federal and state agencies) have formed the lasting iconographic image of the Tar Creek Superfund Lands as a large wasteland, a meaningless ruin of vast dimensions. Anyone looking at these images, which have been exhibited around the country in art galleries and design school venues, cannot help but be moved by the sheer scale of this disturbed place, the two hundred foot high mountains of chat scoured by the wind to form craggy peaks and ridges and valleys, the seventy-five million tons of sandy flats of chat with webs of ephemeral streams and erosion ditches cutting across them, and the eight hundred acres of flotation ponds, red stained with mine tailings. Interspersed among these formations are the communities of five former mining towns, their streets, homes, schools, stores, and people.

Figure 9.2. Collapsed sinkholes, Ottawa County, Oklahoma.

Nancy has also driven along the outlying roads, stood beside tree-lined stream beds, walked up the shifting mounds of chat, looked into the modest yards littered with children's swings and sandboxes, and focused her camera on the playgrounds, ball-fields, parking lots, and streets of the five former mining towns of Picher, Cardin, Quapaw, Commerce, and North Miami as shown in Figure 9.3.

These are two differing sets of views of the Tar Creek site. One presents a broad and expansive lens (so to speak) toward the Superfund lands shaped by large tracts of post-industrial mining land, understood through USGS maps, satellite images, GPS coordinates, production statistics, and historic maps, overlain throughout with small clusters of dwellings, roads, and named communities, and more recently described through medical data of children and teenagers, exposure levels, and remediation reporting.

The other sees, at a detailed scale, the specifics of the daily life of communities – school, work, vegetable gardens, tribal meetings, recreation, church socials, health centers, and baseball leagues, all set within the ravaged environment of Tar Creek (see Figure 9.4).

Figure 9.3. Modest yards littered with children's swings and sandboxes.

Figure 9.4. Daily life of a community: School, work, and vegetable gardens.

272   *Reclaiming the Land*

DON'T PLAY ON CHAT PILES
Bumper Sticker, Paid for by Funds from ATSDR (OSDH 2000)

Attempting to describe the magnitude of the environmental problems in this part of the Tri-state Mining Area is difficult (see Figure 9.5). They include the acidic water in the Tar Creek and the piles of mining waste, the cave-ins, sinkholes, and flooded shafts where men were lowered into the ground to retrieve ore. Yet as Wendell Berry has stated about images such as these, "anybody who troubles to identify in the pictures the things that are readily identifiable, whether roads or buildings, will see nothing in them that is abstract. The power of these places is in their terrifying particularity [as shown in Figure 9.5]. They are the ways, the products and results of human work" (1997: 3).

Figure 9.5. The red-orange acidic waters of Tar Creek.

DON'T PLAY OR FISH IN TAR CREEK
Bumper Sticker, Paid for by Funds from ATSDR (OSDH 2000)

The evolution and reclamation of these places will also be the product of local human work redirected to revitalizing land and restoring the community.

The subject of reclaiming and recycling contaminated land has grown in importance to exert a fascination today that is not fully explained by the influence it has had on the planning and rebuilding of regions, communities, infrastructure, and open space. Revitalizing land and restoring communities requires multiple views and ways of addressing the issues of remediation, reclamation, and recovery – from the broad perspective of planning to the detailed issues of local community development and site design. Superfund sites such as Tar Creek offer three possible models for looking at the landscape and understanding these vast areas of modern destruction.

THREE MODELS OF PLANNING

The first model is simply descriptive and sees a Superfund site as only defined and created by the residue of its previous industrial activities – these include elements such as the waste piles and abandoned mine shafts that are found at Tar Creek.

The second identifies a combination of engineering and design practices in the "technological circumstances of sites," where the actual degraded physical conditions are literally "constructed" or "reconstructed." Here, soil, water-bodies, and landforms are remade and formed through the integration of applied science, engineering, and construction. This is one of the normative ways in which the reclamation of sites such as Tar Creek is conceived and actions are organized and carried out.

The third model, most significantly for planners and designers as well as local community and environmental groups, challenges the traditional models of site planning and design, where "greenfield" or non-polluted sites are understood through consideration of their inherent natural properties, including information regarding geology, soils, vegetation, microclimate, and wildlife. These models are of little use within the contemporary polluted site, particularly in places such as Tar Creek that have been altered far beyond their initial state. The third model offers a new approach for contaminated sites, where the *actual* physical conditions, whether disturbed or contaminated, are identified and addressed as central to the genesis and understanding of the place.

Over the next several decades, federal, state, and local governments will continue to commit resources annually to clean up lands such as Tar Creek that are contaminated with hazardous waste and toxic materials. A number of questions arise from these actions:

- How will large-scale sites shape the national landscape and our future communities, towns, and regions, both within the Superfund sites and outside?
- How do these sites structure the changing needs of scientific, cultural, and aesthetic knowledge?
- How does the adaptation of such elements of the landscape as abandoned mining areas, derelict chat fields, and toxic river corridors affect how the public will perceive and interact with the natural and manmade world during the rest of the 21st century?

DEFINITIONS OF CHAT

The somewhat unusual title of this chapter, "Chat," refers to four definitions and interpretations of the term as well as a number of interconnected themes related to the Tar Creek Site. Chat is a material entity, an economic commodity, a visible and shaped historical landscape, and a community structure that is rarely addressed within traditional models of planning (models 1 and 2, above) and normal federal response and recovery efforts.

CHAT is the outward physical material face of contamination. Granular mine tailings locally known on site as "chat" (a silt-like to pebble-shaped chert and a faceted flint-like rock of remarkable strength and toxicity, containing lead and zinc) are found in large quantities (about 75 million tons, to be precise) on the top of the Tar Creek Site, as shown in Figure 9.6.

Figure 9.6. Detail of chat material.

CHAT is an economic commodity that supported local companies through its commercial use as an aggregate in asphalt, a deicing treatment when mixed with salt, a driving surface layer on gravel roads, and sub-grade material for road construction, playgrounds, and sports fields. In these ways, chat has spread unwittingly throughout the region and over into neighboring states.

CHAT is part of the visible historic industrial landscape, a cultural location of steep-sided man-made mountains – a manufactured geography, regional in scope and an important symbol of past uses of the site – and an artifact of industrial extraction, production, and waste storage, as shown in Figure 9.7.

Figure 9.7. Chat piles, Ottawa County, Oklahoma.

Finally, CHAT can be seen as a community structure dedicated to change over the long term – Communities Helping Achieve Transformations (CHAT). Local environmental organizations in Ottawa County, such as LEAD (Local Environmental Action Demanded) and the Miami High School Cherokee Volunteers, have been extremely effective in highlighting the influence of past mining activities on the current population and concentrating efforts at the community level on informing, advocating, and actively pursuing environmental remedies that enhance the quality of life for families living within the Tar Creek Superfund Site. Here, the sites of recovery and reclamation and the communities and towns are interchangeable. The Superfund site is not in the community's backyards; it *is*, among other things, their backyards – a true case of TIMBY (This Is My Back Yard)!

This chapter explores proposals for the long-term reclamation of the forty square mile Tar Creek Superfund Site through sustainable planning and design activities and strategies generated at the community level. It addresses the ecologies of the area and why they may be considered "resilient." It suggests that while the vastness of the land area of Tar Creek under discussion (four miles by ten miles) calls for a broad regional approach to future planning and reclamation efforts, the area's ecologies have to date represented a failure of the many agencies involved to address decades of environmental degradation, toxic land conditions, and environmental justice issues. I do not suggest that these efforts are in themselves wrong or represent an inappropriate application of agency resources or managerial direction and oversight. Rather, I propose that other models exist for addressing these classes of sites that engage in a more holistic way the true conditions of the landscape in its polluted form with the aspirations and strengths of the local environment and communities.

Local communities and environmental officers, in collaboration with research groups, have begun to address the long-term remediation, planning, and regeneration of vast mining soil heaps, polluted waterways and rivers, and land subsidence and mineshaft sinkholes. For example, with funding secured by Senator James Inhofe (OK), the U.S. Army Corps of Engineers is closing dangerous open mineshafts, mitigating hazards to local residents with land remediation projects, and leading a multi-agency team of experts to assess the potential for subsidence in higher population areas and along major traffic corridors. The partnership includes the U.S. Army Corps of Engineers, U.S. Environmental Protection Agency, U.S. Department of Interior, the State of Oklahoma, and tribal governments, all of whom were encouraged by Senator Inhofe to sign a Memorandum of Understanding in May 2003 to develop solutions to the human health and environmental threats posed by the Tar Creek Site. This led to the Tar Creek and Lower Spring River Watershed Management Plan, a guide for a multi-disciplinary solution for all agencies working to solve the complex environmental problems at Tar Creek.

I will argue, however, that while approaches such as the Tar Creek and Lower Spring River Watershed Management Plan act to initiate transformations of the site, other long-term frameworks and approaches are required to unite community desires, environmental cleanup, and landscape regeneration. This need for a locally based long-range framework indicates not just a change to the physical appearance of land or a simple return to productive use of vast exhausted and currently undervalued areas of ground – a tidying up of the past industrial environment – it signals a profound shift in the way communities must lay claim to this, their own, but very polluted land.

First, however, some basic facts are in order about Tar Creek, the conditions that now exist, and how they came to be.

CONTEXT OF THE TAR CREEK SUPERFUND SITE

The Tar Creek Site is a part of the former Tri-State Mining Area that extended over 700 square miles and included northeastern Oklahoma, southeastern Kansas and reached all the way into the counties of Jasper, Newton, and McDonald in southwestern Missouri. The Tar Creek study area of 2,540 acres over 40 square miles is located above 300 miles of abandoned mining tunnels in Ottawa County in the northeast corner of Oklahoma known as the Old Picher Field lead and zinc mining area. This four-by-ten mile area encompasses five towns, including Picher and Cardin, as well as the sites of over 200 former lead mining operations and milling factories.

Minerals and oil were found in Oklahoma in 1859, with the first commercial well striking oil in 1897. Mining activities were carried out in the Tri-State Mining Area between 1900 and 1970, with operations starting in the Tar Creek region in the early 1900s in an area located above the Boone Formation, an aquifer which provides drinking water for current residents. Mining expanded up to World War II, and between 1907 and 1947, the Tri-State Mining District produced 21.7 million tons of zinc and 18.7 million tons of lead, with a value of more than $1 billion (see Figure 9.8).

At one time, the mountains of milky-white mine waste, as shown in Figure 9.8, proudly proclaimed Picher as the lead mining capital of the United States. After the war, metals demand declined, and by the 1960's mining operations slowed and finally ceased in the early 1970's when companies left the area and the pumps finally ceased to operate in the mines.

Figure 9.8. Mining exploration and waste production.

As early as 1940, the U.S. Bureau of Mines predicted a calamity resulting from the many interconnected tunnels and mine openings filling with water that had previously been pumped out from the aquifer. During the mining operations, to prevent flooding within the working mine faces, acidic water had been pumped from the mines at approximately nine million gallons a day into holding ponds and Tar Creek itself. It turned the stream red-orange, killed the fish and aquatic life, and drove away the beaver and muskrat.

After mining operations stopped, the pumps were turned off and the mines slowly filled with water. Iron pyrite left in waste piles in the mines as well as on the chamber walls and ceilings reacted chemically with the rising water to produce a more potent cistern of acidic water and eventually a 10-billion-gallon vat of subterranean poison, that found its way to the surface as the now familiar acid mine drainage colored a vivid red-orange. In 1978, the U.S. Geological Survey reported that the mines contained 100,000 acre-feet of water. Of that total, the Geological Survey said, 33,000 acre-feet was acidic, which totaled 10,753,097,000 gallons of acid mine drainage water that would eventually discharge into Tar Creek. From there, the acid water spilled into the Neosho River. At their confluence, fishermen found particularly high concentrations of lead, zinc, and cadmium in carp and red ear sunfish. By 1979 water levels were so high that the contaminated water (containing sulfides, cadmium, and arsenic metals) spilled into the waterways, contaminating not only the ground but also leaching heavy metals into the aquifer.

In addition, the miners left behind 1,064 mineshafts, 500 of them open hazards, as well as 10,000 bore holes that littered the landscape in Picher Field and twenty-five wells per square mile that reached down into the Roubidoux aquifer. Approximately 50 million tons of chat or waste mine tailings remaining from the mining operations were left at the Tar Creek Site over an area of 1,200 acres. Of the leftover tailings, 60% has been used to date as road ballast because of its hardness and in driveways, playgrounds, concrete foundations, and even children's sandboxes. These mine tailings, deposited in hundreds of piles and ponds at the Superfund Site, contain lead and other heavy metals. Some of the tailings piles, with residential communities located among them, still approach heights of 200 feet.

The Environmental Protection Agency (EPA) added Tar Creek to the National Priority List in 1981 and has spent more than $40 million dollars to address some of the most disturbing and confounding water, soil, and airborne contamination problems in the country. It now believes the entire cleanup will cost $500 million dollars, a sum few people are prepared to spend on this site.

The silt–sized mining tailings, or chat, as they are known locally, contain heavy metals, particularly lead and cadmium, and pose serious health risks to children 6 years old and younger. Exposure to lead and zinc found in the chat can cause nervous system and kidney damage, learning disabilities, attention deficit disorder, and lower intelligence. A health study in 1999 by the Oklahoma State Department of Health revealed that 38.3% of the children tested in Picher and 62.5 % of the children tested in Cardin had elevated blood lead levels over 10 ug/dL, compared with less than 2.4% for the rest of the state. The expected value for the United States at large is 4.4%. Although 70% of the study area is tribal land, and 42.5% of the population is Native American, the percentage of children considered at risk is not associated with ethnicity.

The fear of contamination has given way to a calm acceptance of the site by the inhabitants. Toxic waste has set the parameters for future action. Here, it is not a matter of demanding a return to pristine nature on the site. Nor is the central issue the use of reclamation as a point of departure for criticism of the site's current state. The fundamental ecological base of the landscape and communities has to be restored.

Large political and social forces threaten some parts of the Tar Creek area, while others seem to have been bypassed by contemporary concerns entirely. Many aspects of the site's past activities remain remarkably intact. All deserve our care and attention, since they, together with thousands of sites large and small, hold the key to the future diversity of human space and what are allowable forms of community and natural life. The surviving towns and communities are devastated by ill health (particularly in infants and children), lack of employment, and the widespread nature of contamination in their local landscape. Yet through all of this, they retain a pattern of strong cultural and regional identity and a spiritual connection to the land and its inhabitants, creatures, and plants expressed through festivals, art, and writing.

> *My opinion of our tour through the Tar Creek Superfund Site is now one of respect for the engineers who are trying to clean up the area. When I first learned about what the EPA was doing, I had very major doubts. It seemed hopeless to try to restore such awful destruction. After seeing it for myself, how they were trying to remediate the problems, I found a lost sense of trust and respect.*

Kit Garvin, 10th Grade
Miami High School, Miami, Ottawa County, Oklahoma (CYC 1999)

In the following section, an aspect of the federal response concerning the soils surrounding residential yards and community facilities is discussed. This is only a small part of a long history of engagement of the EPA, the United States Department of Interior, the Oklahoma Department of Environmental Quality (ODEQ), and the Inter-Tribal Environmental Council of Oklahoma (ITEC) with the Tar Creek Site area. There exists a large amount of documentation of the Remedial Investigation (RI) and Feasibility Study (FI) and subsequent updates and reporting.

EPA RESPONSE

Soil Removal

In 1979, the EPA and the State of Oklahoma became concerned about the acid water and contamination of soils as well as surface and groundwater. The Oklahoma Governor established the Tar Creek Task Force in 1980 to investigate the acid mine drainage. The site was proposed to the National Priorities List (NPL) in 1981 and added in 1983. By June 1984, evaluations had established such a potentially serious risk to human health that a Record of Decision was signed under the Superfund Law. Fourteen years later, in 1998, an EPA Region 6 Emergency Response Team launched a residential soil-sampling program. Soil test results showed that 65% of the homes had yard concentrations greater than 500 ppm. The tables below from EPA documents illustrate property categories within the five towns in the Tar Creek Superfund Study Site – Picher, Cardin, Quapaw, Miami, and Commerce – at the beginning of 2000.

Residential remediation activities have been underway since 1996, targeting first those homes with young children and soil lead concentrations above 1,500 ppm. Soil is removed in 6-inch layers as shown in Figure 9.9 until 18-inch depths are reached or the concentrations of lead are below 500ppm. Clean soil is replaced along with shrubs originally removed at the beginning of the excavation, and new grass, as shown in Figure 9.10.

Figure 9.9. Removal of top 18 inches in residential yard.

Table 9.1. Property categories by town, February 2000.

	Picher	Cardin	Quapaw	Miami	Commerce	Total
Inhabitable Res + 1500ppm	82	4	11	1	79	177
Inhabitable Res + 500ppm	70	-	107	3	251	431
Uninhabitable + 1500ppm	17	1	1	-	1	20
Uninhabitable + 500ppm	11	1	14	-	15	41
Vacant Lot +1500ppm	30	4	1	-	5	40
Vacant Lot +500ppm	21	5	4	5	15	50
Clean below 500ppm	2	-	119	16	136	273
Remediated	504	123	234	160	342	1,363
Denied or final notice	64	12	118	31	223	448
Commercial or Churches	10	1	12	4	11	38
Needs Assessment	12	-	4	-	24	40
Vacant (not assessed)	20	4	2	-	5	31
Total	**843**	**155**	**627**	**220**	**1,108**	**2,953**

Source: USEPA

Table 9.2. Yard remedial action status, March 2000.

Cost of Work Performed	$22.2 Million
Property (Yards) Completed	1124
Total Cost per Property	$19,800.00
ROD Estimated Cost per Property	$20,000.00
Number of Cubic Yards Excavated	284,000
Dollars per Cubic Yard	$78

Source: USEPA

Figure 9.10. Completion of soil removal with replacement soil, turf, and shrubs.

The benefits of federal agency actions as reported in April 2005 are as follows:

The cleanup of lead-contaminated soils from 2,071 residential yards and high access public areas located within the five-city mining area has significantly reduced the exposure of the population, especially young children. Recent independent studies comparing blood lead level data collected in 1997 to data from 2000 show an approximately 50% decrease in the number of children living in Picher and Cardin between the ages of one and five years old with blood lead levels to or greater than the 10 μg/dL standards set by the CDC. A report in 2004 by the Agency for Toxic Substances and Disease Registry (ATSDR) shows that children living at the Tar Creek site who had blood lead levels in excess of the 10 μg/dL level decreased from 31.2% in 1996 to 2.8% in 2003. This reduction in the number of children with elevated blood levels is attributed to the residential yard cleanups and extensive educational efforts by federal, state, county, and tribal entities. Abandoned well plugging has also reduced the potential for contaminants in the shallow Boone Aquifer to migrate to the Roubidoux drinking water aquifer (EPA 2005).

The EPA's program of yard removal is almost worthless. Without public education, the repaired properties will become contaminated again quickly.

Chris Robinson, 10th Grade
Miami High School, Miami, Ottawa County, Oklahoma (CYC 1999)

As Earl Hatley, former environmental director for the Quapaw Tribe of the Tar Creek region notes, the winds will continue to blow across the chat piles, and two years after remediation the yards will be re-contaminated to the same levels or higher. No further testing has been carried out to confirm or disprove his theory. However, it is likely that the windborne chat will continue to be distributed from the piles across the Tar Creek study site in a microscopic granular layer. In addition, in some cases, "clean soil" has meant water-impermeable clay. The remedial work that removed contamination left the new yards full of clay, homes are retaining moisture from rains, and mold has developed at an alarming rate. This is particularly distressing given the already substantial incidence of asthma in the area.

I want to return to the central argument of this paper regarding the contrast between the approaches (planning models 1 and 2) used by the state and federal agencies to address polluted sites such as Tar Creek and the local knowledge and understanding the *actual* physical conditions, whether disturbed or contaminated, that can lead to an alternative (model 3). This model is more sustainable for the long-term approach of cleaning up and regenerating large and complex land areas over time and including the close involvement of all stakeholders including community partners and participants. The difference between the two sets of approaches is striking. On the one hand, the normative methods (models 1 and 2) used under current modes of engineering analysis, evaluation, and response focus on a narrow way of looking at contamination *in situ*. In short, the site exists in an arrested state of succession or development where the levels of contamination can be measured, evaluated, and removed within a single timeframe. The site is then returned to a clean state, a *tabula rasa* that is devoid of any pollutants, a natural benign landscape where the land, dwellings, and people are in harmony. On the other hand, the above process appears to have solved the most pressing need – that of immediately addressing the health problems of lead poisoning faced by the town's residential communities, particularly the young and the elderly. By removing the immediate source of lead in the top eighteen inches of residential yard soil and replacing it with clean soil, an immediate result is gained. It can be quantified, measured, and mapped and to date has resulted in the most serious efforts in economic and engineering terms to remediate the Tar Creek area.

While the approach embodied in planning models 1 and 2 looks to a short-term response to "solve" the problem of contamination, it avoids what turns out to be the true difficulty associated with sites such as Tar Creek –

simply that they are the product of a relentless industrial process (mining) that occurred over a long period of time. In doing so, the mining activities and their residues and wastes acted gradually on the environment to alter ecosystems, adjusting the nature of hydrologic systems and the structure of soils and sediments. Vegetation, wildlife, and aquatic life were also severely altered. Finally, the interaction of these patterns of contamination and alteration of the soils and water with the human population has occurred and will continue over a significant period of time, even though the soil removal exercise from residential yards has taken place. It is not possible to undo over a hundred years' worth of mineral extraction and its large-scale effects on the land in an instant, nor is it possible to continue to view the process of remediation as a simplistic linear activity. Rather, it requires an adaptive model that accepts the changing circumstances of pollution and remediation techniques over time, in concert with a profound shift toward communities and local stakeholders being involved as active participants rather than passive onlookers.

This speaks to the inability of models 1 and 2 to answer the deeper and more significant questions associated with the Tar Creek site. The environmental disturbance is vast and widespread and will require long-term remediation efforts to address the chat piles, the condition of the water bodies, the subsidence of the ground, and the airborne contamination. It also speaks of the need to move toward a different way of addressing the planning of the reclamation of the Tar Creek Site area, one that is more receptive to the actual conditions of the site, both physical and cultural, as well as to involve all stakeholders and communities in the process. Large-scale long-term remediation requires local knowledge and a large-scale design vision to be attained in small stages over significant periods of time. Some of the challenges posed by these large-scale remediation efforts are immense, concerned with incremental changes over time to the very nature of the landscape, continuous monitoring of both the original conditions and the remediation in progress, and adequate mapping of the cultural as well as environmental fabric that is to be identified and addressed as central to the genesis and understanding of the place.

In direct contrast to the EPA's selected remedy – removing the top eighteen inches of soil in residential yards, taking it away to be stockpiled in a lagoon, and replacing it with clean topsoil, reducing the amount of lead in the soil to below 500 parts per million (ppm) (Alternative 2 in the March 1997 Proposed Plan) – one part of the study project that follows tests a hypothesis that addresses the key points expressed by Earl Hatley on the long-term nature and extent of contamination at the site. It further seeks to demonstrate that community-based planning models can more effectively curtail high soil lead concentrations in targeted study areas within the Superfund Site. Efforts are underway through this project to integrate a locally-based site strategy with a long-term land renewal effort.

THE HARVARD COMMUNITY-BASED PREVENTION AND INTERVENTION RESEARCH PROJECT

The Tar Creek research project currently underway combines the efforts of faculty and researchers across three graduate schools at Harvard University – the Medical School, School of Public Health, and Graduate School of Design. Founded through contacts that arose out of various subcommittees across the University that were engaged in developing interdisciplinary faculty research on the environment, it involves Dr. Howard Hu of the Harvard Medical School, Dr. Lucia Lovinger of the Harvard Science Center, Professor Jack Spengler of the Harvard School of Public Health, Dr. Robert Wright of Brigham and Women's Hospital in Boston, and the author, representing the Graduate School of Design, as well as Rebecca Jim of L.E.A.D, the local community group on site in Oklahoma.

The Harvard Community-Based Prevention and Intervention Research Project has three main components. First, the project includes a community-based study of lead exposure biomarkers focusing on children in the study area. Second, there is a study of lead-exposure pathways tracking the movement of lead from mine spoil to various populations that includes studies on cultivated and foraged foods consumed by residents and wildlife. Third, in the future the project will examine remediation design tools and approaches to reduce lead at the larger site scale through innovative remediation and design techniques. It proposes to develop a model that will join other efforts at environmental remediation of contaminated areas, such as a pilot passive wetland treatment system and a wetlands and wildlife refuge, while addressing land subsidence, flooding, mine drainage discharges, and chat pile distribution. For example, in its most vivid approach to community-based remediation, it will attempt to join efforts at environmental remediation of contaminated sites using living planted systems, an emerging technology called phytoremediation, with long-term environmental restoration and reclamation of the broader land and water bodies. Thus, it promises to integrate a locally-based site strategy with a long-term land renewal effort and supports the necessity of moving to an adaptive management approach (model 3) to address the future of Tar Creek in the first half of the 21st century.

At this point in the paper, a description of the model proposed is offered with reference to its application to the Tar Creek Site. In conventional approaches to remediation, the act of cleaning up a polluted site is viewed as a set of extremely particular and separate engineering actions. These actions or cleanup methods lead to a final result that is measured and tabulated in relation to their subsequent effect on the soils or water on a site. These media are, in short, returned to a less polluted state. While this fulfills the original purpose of the conventional remediation process, such methods do not relate to, nor are they integrated with, other natural and man-made processes on the site (or each other for that matter). Nor do they assist in spatially organizing or structuring long-term reuse of the site, or in setting up a process for determining how the site may be further healed over time in terms of its sustainability for natural systems or human populations. In addition, they do not relate to possible cultural or social activities that may occur as part of the community. They happen quite simply in a vacuum and usually securely behind a fence.

The adaptive model proposed takes a very different starting point. It assumes that remediation will be a long and ever-changing process. The model will be required to adapt to alterations of landscape conditions – soils, vegetation, water quality, land patterns and uses, and microclimate – alterations that occur as a result of early remediation efforts. Most importantly, the process of land regeneration has to be integrated with the local communities that occupy it. The model has it roots in the ecological process of succession. Like a forest that starts from a ploughed field or burnt stubble from a forest fire, a series of shifting mosaics of plants and communities colonize the ground. In time, they afford shade, shelter, and food, and give other plants and species habitat for survival and subsequent growth. Nothing goes according to a perfectly scripted plan, and there are always obstacles to overcome such as wet soils, depressions, rocks, or lack of sunlight. The landscape in time can be understood as a series of small local adjustments to the particulars of the found environment. So it is with long-term plant-based remediation approaches. The Tar Creek Site is such a place of initial devastation like the ploughed field or burnt stubble, with even more complex problems to overcome at the outset. In order to start a comprehensive process of Superfund succession, attention has to be paid to existing site conditions, not just the location of polluted areas and their characterization but other factors including natural systems of vegetation, hydrologic systems, wind patterns, microclimate, wildlife habitat, and human patterns of settlement, land use, and transportation. Finally, care has to be given to cultural patterns of use in the site area, including those that have a sacred or symbolic character for the inhabitants. In the case of Tar Creek, the river system itself plays a significant role in the community, both as symbol of the devastation brought to this place, and as a future medium of cleansing and hope. The ironically named Tar Creek Fishing Tournament, held annually and sponsored by LEAD, uses the event to draw people's attention to the ability of the Tar Creek to once again exist as a vital part of the community's landscape.

INNOVATIVE COMMUNITY-BASED REMEDIATION

One important aspect of the Harvard Collaborative approach is the use of a localized remediation strategy – phytoremediation – to engage the local community in planting, monitoring, and harvesting, as necessary, that works the land in a respectful way that more closely mirrors agricultural practices than engineering remediation. The continuation of the community and local stakeholder investment in this activity is to persist over time as the long-term evolution of the phytoremediation system works to both clean up pollutants as well as establish a new plant ecosystem that will vegetate the chat piles, clean up wetlands, and clean residential yards in an organic and holistic way. The extended timeline associated with this process is usually seen as lasting over thirty years, with short-term results occurring within one year or three growing seasons.

Drawing on basic research about the natural abilities of plants to accumulate lead from soils and groundwater, phytoremediation (literally meaning plant-cleanup) is viewed as a low-cost alternative natural treatment technology for removing lead contaminants. It is an *in situ* technology that uses plants, including trees, grasses, and aquatic vegetation, to remove, sequester, and uptake hazardous substances from the environment. It offers promise as a versatile strategy suitable for use on contaminated sites with a range of pollutants.

The term "phytoremediation" was coined by Dr. Illya Raskin of Rutgers University in 1991, although the process itself had been in use for thousands of years as civilizations used plants to uptake and treat water. Early exploratory studies were carried out, primarily in Europe, mainly related to the reuse of mining spoil and waste sites in the rebuilding efforts following World War II. Research, however, did not start in earnest until the mid-1980's in the United States, when the original mechanisms of phytoremediation were described in laboratory experiments. These can be grouped into three broad categories, based on the fate of the contaminant in question: accumulation, degradation, and hydraulic control. Optimizing the effects of phytoremediation, the plants are generally densely planted in either rows perpendicular to the flow of a subsurface groundwater plume or in clumps or patches known as hotspots where the contamination occurs in concentrated soil areas. These have the ability to be applied in areas of high lead within communities including playing fields, river walkways, and parks. The question posed by such a technology regards the increased net environmental benefit of integrating a community-based model phytoremediation plan to reduce lead levels with other community concerns, and carrying out such efforts over the long term.

Experience with phytoremediation in the field suggests that the adaptive model is ideally suited to long-term strategies that are successional, that is, they will undergo adaptive change as part of their evolution. An example can be found in an initial phytoremediation installation that has to address a particular condition in the soils or groundwater. The contamination is addressed over time as the plants that are installed grow and flourish. In time, they form habitat for wildlife and begin to establish a dense forest edge near to a recreational area. At a certain point, a portion of the forest is harvested for local timber and the remaining trees mature continuing to uptake contaminants, some of which have only recently found their way into substrate. Certain trees may become diseased and die; others may be felled by lightning or winds. In this, the process proceeds from its starting point (the point of installation), altering and adjusting over extended periods of time.

A Single Bracelet Does Not Jingle

Rebecca Jim has often stated in presentations an old proverb that says "a single bracelet does not jingle." Phytoremediation research projects by their very nature foster collaboration between different fields of science and between different professions and communities. The potential also exists to integrate phytoremediation with other forms of environmental planning and design enterprises: for example, pump and treatment wetland systems, greenways, and solar technologies. The cleanup of lead by phytoremediation has been shown at a number of full-scale demonstration projects as an alternative to more established treatment methods used at hazardous waste sites, such as soil removal and replacement as has been carried out to date at Tar Creek. Limitations to its application at more conventional sites include the extended time-scale of phytoremediation activities in comparison to competing remedial technologies, and the inherent limitations of biological systems. Its low cost (approximately 70-100 times cheaper than conventional remediation techniques) makes it appropriate for sites such as Tar Creek, where a longer time-scale is necessary and the site holds the potential to integrate this approach with other community-based land redevelopment projects.

Initial testing of a phytoremediation approach using vegetation in the field was carried out over the summer of 1999 by Harvard Design School graduate student Markley Bavinger. Through all phases of the research, adults and college and high school students living in the community were involved in the survey of opportunistic and/or native vegetation and their patterns of growth in targeted site areas. Design projects will later address specific test areas where phytoremediation installations will be integrated with community proposals for the river corridor, the containment of the mining waste sites, and residential areas and school yards. Phytoremediation is effective only in certain conditions. A phytoremediation scheme will make sense only if there are appropriate growing conditions, contaminant densities, soil aeration conditions, and time constraints on remediation. Planners, scientists, and communities can play a role in the prevention of polluted landscapes by designing site and land-use codes that anticipate the presence of toxins and mitigate that presence with hyperaccumulators. During the "Manufactured Sites" Conference at the Harvard Graduate School of Design in 1998, one participant explained, "On some sites it is possible to place planting in such a way as to allow partial reuse of the site for public access or ongoing development while the cleanup is in process. Here, phytoremediation and creative site design are united by the use of planted systems that both remediate and at the same time establish spatial and functional patterns of use."

Some of the preliminary efforts to address these issues within the Harvard Collaborative study have resulted in an examination with local environmental groups of how the labor-intensive remediation planning and installation process could be integrated into the land ethic projected by various community organizations. This process involves educational initiatives by Native Americans to encourage various groups to become engaged in the monitoring of ecosystems, as well as to review agricultural practices in the area, particularly where food production is coupled with remediation. The issues of coordination and management represented by model 3, the adaptive management approach, are enacted at the local level where existing native vegetation has been inventoried and mapped alongside known phytoremediation vegetation that can address local pollution. This, in turn, has led to examination by local environmental groups and members of the Harvard Team of the patterns of vegetable and crop production in the area and the ability of these planted systems to act as further exposure pathways for contaminants in the community (i.e., through diet).

Governor Frank Keating's Tar Creek Superfund Task Force

Parallel to the work by the EPA and the Harvard Collaborative Group, a report on a major initiative on Tar Creek was issued in October 2000. Prepared by the Oklahoma State Office of the Secretary of Environment, it was commissioned by the Governor to give further impetus to the resolution of problems that remained at the Tar Creek site. The vision for the site was:

> *To establish a world-class wetlands area and wildlife refuge within the boundaries of the Tar Creek Superfund Site that will serve as an ecological solution to the majority of the most pressing health, safety, environmental and aesthetic concerns* (Executive Summary 2000).

This study, in addition to the principles underlying the Harvard project, suggests an awakening of interest in using a sustainable long-term approach to the problems of Tar Creek and its communities.

The Interim Measures established by the Final Report include the following, to be carried out as a Superfund Remedial Investigation/Feasibility Study (RI/FS):

- Undertake a comprehensive study of mine-waste drainage to determine amount and types of treatment wetlands;
- Initiate a pilot wetland treatment system to determine the most effective and feasible systems for the area;
- Develop and test appropriate uses of large volumes of chat to establish markets for its export;
- Identify mine shafts that need plugging and test various closure methods; and
- Investigate the bio-availability of heavy metals resulting from the consumption of wild forage foods.

In addition, the following tasks are to be carried out:

- Establish a local industrial authority to develop the necessary infrastructure and appropriate regulatory climate for large-scale aggressive chat marketing and export;
- Assemble a local steering committee to explore options for, and feasibility of, the relocation of the towns of Picher/Cardin;
- Create a Geographic Information System (GIS) *ad hoc* committee to compile all available data and information on the Superfund area, to allow for better decisions to be made about all aspects of the future remediation effort; and
- Develop a better EPA, DOI, Tribal, and State partnership to hasten cost recovery /Natural Resource Damage Assessment reimbursement.

$22.2 million has been spent over almost twenty years to address some of the most confounding water, soil, and airborne contamination problems in the country. The remediation and recovery as currently defined by the Governor's Report should cost $40 million dollars, a sum few people are prepared to spend on this site although the scope of remediation is still limited. However, the integration of a larger living ecological system – the wetland – with more locally-based initiatives as outlined in the Harvard study suggests that steps can be taken that can involve not only remediation but also restoration of a broader ecosystem, which would bring in partners and funding from a wider range of agencies such as the Department of Interior and Fish and Wildlife Services as well as the Audubon Society and other environmental groups.

ENDINGS

How can stakeholders in the Superfund process, whether planners, lawyers, designers, or community leaders, reinterpret the chronology of a site, be it natural, industrial, or cultural in its disposition, in a specific geographic location? And how can reclamation and reuse activities shape a Superfund site into a community environment within the cultural artifice of a post-industrial landscape? The Harvard Collaborative study is focused on carrying out a large-scale plan that accommodates and integrates local communities in a long-term remediation of the land and waterways, using a plant-based technique that acknowledges the specifics of the ecology and physical nature of the local environment while creating a permanent infrastructure of vegetation across the landscape, a tapestry not of chat piles and sink holes, orange colored ponds and desolate ground. In this plan, plants native to the area assist in the absorption and elimination of pollutants, guided by communities that share the site and the cultural practices of the place. In addition, these plants, used in broad fields, boundary patches, and avenues, begin to form a living blanket of trees, shrubs, and ground covers at a scale not of a residential garden but rather a regional grassland park.

We have found that, with reference to the Tar Creek Project, putting the site within a broader cultural and ecological context has considerable implications for reuse options for this grassland park concept used in selected parts of the area and for how further remediation selection has to proceed in the future. In addition, the community has to date engaged incrementally with the residue of mining activities and historic traces as a way of reclaiming their own territory and past, looking at the remediation as a transparent and legible process involving all members of the community. Finally, viewing the site over the long term as an effort from the bottom up is essential to successful project implementation. Success should be measured by small-scale achievements instead of normal benchmarks for reuse – by the flourishing of new green landscapes rather than reductions in pollution.

By looking observantly upon the natural world as well as the disposable world, we may build at the great overlap between the two. This suggests a challenging new model for how we ought to work within the contemporary Superfund environment with a new quality of attention to the intricate organic and artificial systems of reality. The application of adaptive management methods to landscapes that have undergone drastic environmental degradation is only in its infancy. Overcoming the normative methods currently understood by the vast majority of those employed by federal and state authorities will rest in part on the ability of individuals like Rebecca Jim of Tar Creek to slowly and sustainably rebuild the landscape in which she lives.

REFERENCES

Berger, John, and Jean Mohr. 1981. *A Fortunate Man*. London: Writers and Readers Publishing Cooperative.

Berry, Wendell. 1997. "Preface." In David T. Hanson, *Waste Land: Meditations on a Ravaged Landscape*. New York: Aperture.

Concerned Youth and Citizens, Miami, Oklahoma. 1999. "The Legacy." In *Tar Creek Anthology*. Miami, Oklahoma: LEAD (Local Environmental Action Demanded).

CYC. *See* Concerned Youth and Citizens.

EPA. 2005. U.S. Environmental Protection Agency Report, EPA Region 6, Congressional District 02, Tar Creek (Ottawa County). October 4, 2005.

Executive Summary. 2000. Executive Summary, Final Report of Governor Frank Keating's Tar Creek Superfund Task Force. October 1, 2000.

Oklahoma State Department of Health. n.d. Part of a series of bumper stickers on the subject of prevention of contact with chat piles and polluted water courses issued by LEAD (Local Environmental Action Demanded) and paid for by the Agency for Toxic Substances and Disease Registry (ATSDR), Oklahoma State Department of Health.

OSDH. *See* Oklahoma State Department of Health.

CONCLUSION

Gregg P. Macey
University of Virginia Department of Urban and Environmental Planning

Jonathan Z. Cannon
University of Virginia School of Law

As we enter the third generation of Superfund, the program faces some stark realities. One in four Americans continue to live in close proximity to a toxic waste site. The program's taxing authority expired at the close of 1995, and in 2002 the Bush Administration announced that it would not seek reauthorization. These decisions have left the program to cope with a decrease in funding. The number of sites remediated per year declined by roughly 50% from the late 1990's to 2003, continuing a backlog that perpetuates health risks for local residents. And while a growing percentage of sites listed on the NPL (about 60% in October 2005) are considered "construction complete" (all immediate threats addressed and long-term threats under control), many of those sites will demand years or even decades of operation and maintenance and regular monitoring and review for indefinite periods.

Further complicating the operation of Superfund are uncertainties about rules governing liability between and among responsible parties. In its recent decision in *Aviall Services, Inc. v. Cooper Industries*, the Supreme Court held that a PRP who voluntarily assumes cleanup costs is not allowed to use CERCLA's contribution provision to sue other PRPs. This has left lower federal courts struggling to determine whether and under what legal theories a volunteer PRP can force others involved at the site to help fund the cleanup. As uncertainty over who will pay to clean up sites continues and funding generally dwindles, the program faces even costlier and more complex sites. Indeed, more than half of the program's budget is presently allocated to cleanup efforts at *eight* sites.

Facing a new set of challenges, the Superfund program has responded in part by shifting expectations. While the program at first focused on the number of sites that it could remove from the NPL, it redefined success as achieving construction complete status in 1993, acknowledging that such sites would still require long-term, onsite activities. We have argued that the third generation of the Superfund program calls for more explicit attention to adaptive site stewardship and what it entails in the post construction completion phase. Specifically, Cannon argues that the EPA should adopt a flexible and expanded view of the public good that moves beyond the values represented in the statute, promote institutional innovations to clarify and order these values, improve monitoring and feedback mechanisms focused on crucial unknowns at a given site, encourage the integration of decisions across sectors and jurisdictions, and employ conscious policy learning to the entire portfolio of sites in its inventory. As many contributors to *Reclaiming the Land* have noted, the program as it stands is ill-equipped to address these matters of scale, hierarchy, and conscious learning, leading the EPA to adopt thematic elements of adaptive management piecemeal while neglecting to incorporate it into the cleanup process more generally.

Adaptive management fits squarely within a new approach to environmental regulation built on a more "modular" architecture to address the fundamental questions facing regulators: what level of government should manage environmental challenges, and using what tools? Modularism, of which adaptive management is one of several key elements (see Freeman & Farber 2005 for an excellent introduction to the approach), rejects the traditional approach of regulatory agencies that operate independently, with minimal stakeholder input and inflexible regulatory tools, and without the means to adjust to changing conditions. Modularity is built on a foundation of coordination across agencies and different publics, flexible problem-solving, formal and informal tools, and new governance structures that can manage a variety of competing demands over longer time horizons.

Adaptive management provides a participatory dimension that can correct errors through deliberation and dispute resolution, one version of which was sketched by Dukes in his discussion of consensus-building. Building on years of facilitation experience at Superfund sites, he demonstrates how planning for site reuse involves a series of dynamic variables that belie attempts by agencies to apply immutable formulas for community involvement. Dynamic variables include unpredictable timetables between listing and construction completion, changing agency personnel, uncertainty about the nature and extent of contamination, and evolving community preferences. Dukes' challenge to conventional thinking regarding community involvement recasts consensus-building as a tool for learning and error correction within and across the portfolio of Superfund sites. Bluestone adds to the participatory dimension of adaptive management a call for deliberations to address site histories, to encourage more informed community involvement and decisions regarding future uses. As remediation work progresses, innovations in industrial history and historic preservation can help stakeholders to treat cleanup and preservation as compatible goals, adding a new dimension to site stewardship.

As Dukes and Bluestone show how collaborative mechanisms can promote the kind of learning to which adaptive management aspires, new approaches to social interaction beg the question of both the *setting* and *content* of social interaction – how will different levels of government be involved and integrated, what information such as simulation models and historical narratives will be developed and shared, how will tradeoffs and coalitions be built, and how will surprising outcomes be addressed? Cannon, Hernandez, and Landreth address the scale of governance question and institutional barriers to the effective use of adaptive management. Cannon explains how Superfund is a program for contaminated site management that involves multiple parties, over extended time periods, and across a range of values and policy objectives. He criticizes the program's focus on time-limited intervention directed by federal officials focusing on federally defined public health concerns. Specifically, he notes the authority for remedial decisions at NPL sites is given to federal officials and is not delegable to state or local jurisdictions. In practice, this focus is self-limiting: while remedy selection is driven by the federal government, it is property owners and local officials who determine site use, a reality that impacts implementation and effectiveness of a chosen remedy. Adaptive management offers a means of integrating decision-making across scales to more responsibly address how local factors and forces will impact project design and execution over time.

Hernandez and Landreth expand the analysis to brownfield sites, contrasting CERCLA's focus on cleanup as the ultimate goal with the redevelopment focus of market forces, local real estate and land use realities, and available investor capital. They call for an expansion of such EPA notions as "reasonably anticipated future land use" to address local realities, and incorporate them into the remediation process as early as possible. The key to the long-term management of a site, they argue, is to use institutional controls not normally associated with CERCLA and RCRA (proprietary, government, enforcement, and informational) to anticipate both short-term remediation and long-term site prospects. These localized tools can lessen the difficulties posed by monitoring needs at both small- and large-scale contaminated sites while streamlining their management over extended periods of time. The next generation of Superfund will have to address these tools and encourage their use across different scales of governance.

Adaptive management also demands innovation in the tools available for generating information that will be shared, considered, and used to manage contaminated sites. Beling et al. push the envelope with a systems analytic approach that considers a range of beneficial uses available for sites. In order to preserve future options, the authors demonstrate how to avoid dead-end decisions and associated back engineering, which in addition to being incompatible with the central tenets of adaptive management would add substantial costs to site cleanup. Remedial project managers are encouraged to use an adaptive management approach to determine and mitigate the effects of contingencies such as institutional uncertainties, site parameters, future reuse options, and technological availability. The notion of policy-oriented learning is expanded even further by their approach, which can be used to evaluate reuse options at individual sites as well as site portfolios.

Another methodology for generating site information is offered by Kenney and White. They take a standard discounted cost-benefit analysis approach, which is typically used by modelers who begin their analyses "with the end in mind," and tweak it to allow for flexibility and openness to change. They build on the strengths of the model, including its ability to consider tradeoffs among competing uses and technologies by interested stakeholders, ease of model construction, and incorporation of diverse preferences. To these strengths they add model characteristics that will allow remedial project managers to work with other stakeholders, adjusting model parameters and considering different cleanup and reuse options. The most promising element of their work for adaptive site stewardship is their model's ability to address the factors affecting reuse *before* the cleanup phase even begins at a contaminated site. Such a model promotes cost-effectiveness while furthering the participatory goals of adaptive management.

Satterstrom et al. address a third analytic approach to data generation, multi-criteria decision analysis (MCDA). Following an exhaustive review of the acceptance of adaptive management within government agencies in the U.S., Canada, and Europe, they conclude that the quantitative methods needed to implement the approach lack a broader framework for integrating the people, processes, and tools required to make appropriate cleanup and site reuse decisions. An adaptive management paradigm is important, they show, because acceptance of uncertainty at the RI/FS phase leads to better decision-making by remedial project managers in the cleanup phases that follow. But modeling has its limits, and large-scale modeling called for by such tasks as predicting ecosystem response is fraught with error. Thus, the authors call for the addition of optimization loops after initial decisions are made. MCDA can help with this, providing a structured process that is transparent and inclusive: multiple parties can help to define problems, generate alternatives, determine the relative importance of decision-making criteria, and return to previous decisions to make adjustments based on past performance.

To give context to our call for adaptive site stewardship, we ended this volume with a case study of a ravaged landscape from which hopeful signs of the new approach have emerged. Kirkwood carefully details this landscape, which traditional planning models would frame in terms of its "technological circumstances," such as its soil, water, and landforms and the isolated engineering practices that have been built up around each characteristic. EPA's soil sampling and removal program at Tar Creek is instructive. A tenth grader's rebuke of the program points to the limits of the traditional model, which in this case focused on soil lead levels in a small number of targeted yards and areas of public access. Much is sacrificed by an effort that focuses on an immediate result that can be measured and mapped within a single timeframe. Solving a problem in the short-term is often necessary. But Kirkwood suggests that the Superfund program lacks the broad-based involvement, coordination across levels of governance, and analytic tools needed to address the interaction of soil and water contamination with human populations, a process that has altered ecosystems in the Tar Creek area across generations and will continue to do so in ways both straightforward and nonlinear for many years to come (recontamination being just one example of how this will occur). Initial efforts encouraged by a community-based research project provide some promising developments in that direction.

Proponents of adaptive management argue that moving beyond fragmented environmental regulation, with its focus on standard setting and delegation of authority, is a "matter of conscious design" (Freeman & Farber 2005: 799). Contributors to this volume have worked to create new building blocks for use in such an endeavor, whether they are new processes, data generating tools, or modes of cooperation. But more must be done to identify optimal institutional arrangements. Agency officials at all levels of government must work closely with others to identify solutions to the unique problems posed by Superfund sites – particularly the kinds of sites that dominate the program's efforts of late. We hope that the elements of adaptive management will inform that effort, expanding and enriching the conversation and helping shape the outcome.

REFERENCE

Freeman, Jody and Daniel A. Farber. 2005. "Modular environmental regulation." *Duke Law Journal* 54: 795-912.

INDEX